Learning Robotics with ROBOTIS PLAY Systems

By Chi N. Thai

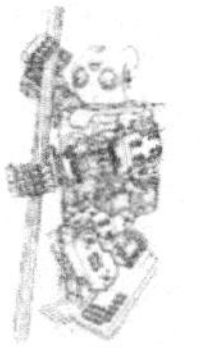

CNT Robotics LLC, Lawrenceville
2017

Chi N. Thai

roboteer@comcast.net

www.cntrobotics.com

Learning Robotics with ROBOTIS PLAY Systems (2017)

ISBN 978-0-9993918-0-8 (Paperback).

Library of Congress Control Number: 2017914710.

Published by CNT Robotics LLC, Lawrenceville, GA, U.S.A.

Printed by CreateSpace.

To my parents for giving me life.

To Christine and Emily for their enduring love and support.

To the next roboteers: Nicholas, Adrian, Madeleine, Josie, Devin.

Preface

The PLAY700™ programmable robotic kit was released by ROBOTIS® in December 2016 for the USA market with a cost under $90 US. It is recommended for users at age 8 or older. It comes with a small assembly manual for four robot examples and it can be used with two free popular programming interfaces:

1) The first interface uses a ROBOTIS tool called TASK™ which can generate machine code that runs on the robot controller CM-50 allowing it to interact with its built-in motors, NIR sensors and miniature speaker. These TASK codes can also link to a mobile device (smartphone or tablet) via **Bluetooth®** to access many mobile services such as audio, video, text/graphics display, text-to-speech, speech recognition, camera, etc… to enhance both user interactions and robot acquisition of features beyond its original actuators and sensors.

2) The second interface uses the off-line version of MIT's SCRATCH® 2 software to combine the power and multimedia services of a Windows® PC with a direct control of the robot controller CM-50 via Bluetooth and a helper application named R+SCRATCH, provided by ROBOTIS.

Who is this book for?

This book for you if you are a young robotics enthusiast ready to go beyond the four original robots WINDMILL, DOG, CAR and SCORPION, or if you are an adult investigating the possible use of the PLAY700 kit to help children learn about robotics programming and design. This book will show that this kit can be quite a versatile tool to introduce students from 8 to 12 years old to fundamental concepts in several areas: computer programming, robot control, inter-device communications and multimedia programming for richer story telling.

What do you need?

At the minimum, you need to purchase the ROBOTIS PLAY700 kit which is widely available via several web stores. You also need to have access to a mobile device (phone or tablet), and have the mobile Apps R+mTASK and R+PLAY700 installed on these devices. Depending on the operating systems on your device, these Apps are available at the appropriate App Stores (https://itunes.apple.com/us/developer/robotis-co-ltd/id948481761) or (https://play.google.com/store/apps/developer?id=ROBOTIS&hl=en). With this configuration, the user can create TASK programs that run on the Hardware Controller CM-50 and interact with the built-in actuators and sensors, as well as access graphics, audio and video facilities on mobile devices via a Bluetooth module named BT-410 (included in the PLAY700 kit).

If the user wants to use the PLAY700 kit with the SCRATCH 2 software, then the user is currently restricted to Windows® PCs, as the needed helper software named R+SCRATCH is currently available for Windows OS only. Consequently, the user needs to use the Windows version of the SCRATCH 2 Offline Editor. This web site (http://www.robotis.us/software/play700/) contains download links for all needed PLAY700 software and tutorial videos. Furthermore, as the furnished BT-410 is not compatible with Windows OS, the user is required to purchase an additional ROBOTIS USB module named "BT410 Dongle" so that the Windows PC can communicate with a PLAY700 robot equipped with a BT-410.

What is in this book?

This book consists of 5 chapters:

1) Chapter 1 presents foundational concepts regarding the system design approach used by ROBOTIS in creating its educational kits and the Sense-Think-Act paradigm used in developing the contents of this book.

2) Chapter 2 describes the hardware characteristics of the PLAY700 kit and how to use the ROBOTIS MANAGER software tool using a "Basic Bot" design.

3) Chapter 3 is a substantial chapter providing a gradual but in-depth tutorial about applications of the R+TASK and R+PLAY700 software tools using two robot designs - "Spinning Top" and "Car Bot". Topics included autonomous-behavior and remote-control algorithms, communication and multimedia programming techniques. PC and Mobile uses of the TASK tool will be shown.

4) Chapter 4 is also another substantial chapter mirroring the instructional approach and topics developed in Chapter 3 but now using the R+SCRATCH/SCRATCH 2 tool chain and its event programming features.

5) Chapter 5 presents selected mechanical design concepts inherent in the mechanical components provided in the PLAY700 kit, with the goal of helping students understand the mechanical design concepts represented in the original example robots, and be creative in their own robot designs by showcasing additional mechanical concepts and robot designs.

Chapters 2 through 5 also provide appropriate source codes and tutorial videos (via YouTube®) to illustrate the presented concepts, along with review questions to help students master learned materials. Please visit www.cntrobotics.com/playbook for download links to the source codes and tutorial videos.

How should you use this book?

All readers should first read Chapters 1 and 2. Ideally, readers should practice the MANAGER related procedures as described in the text of Chapter 2 and its companion videos, but this practice would require the use of the USB BT410 Dongle.

Chapter 3 should be tackled next if readers are interested in using TASK and R+PLAY700, i.e. to have robots that work closely with mobile devices.

Chapter 4 is for users who are interested in using SCRATCH 2 and want to build their expertise in applying event and multimedia programming techniques to robotics.

Chapter 5 can be read independently from all other chapters.

And please, do let the author know about your experiences in using the materials in this book, this will help the author improve the next edition and the companion resources. Please visit the author web site www.cntrobotics.com for updates and other robotics materials. Please send your comments and suggestions to roboteer@comcast.net.

Trademark Attributions

Content in this book made references to the following trademarks:

ROBOTIS product names are trademarks or registered trademarks of ROBOTIS, Inc. Used with permission by ROBOTIS Co. Ltd.

Android®, YouTube®, and Chrome® are registered trademarks of Google Inc.

SCRATCH® is a registered trademark of the Scratch Team at the Massachusetts Institute of Technology.

Windows® is a registered trademark of Microsoft Inc.

iOS® and iPhone® are registered trademarks of Apple Corporation.

Bluetooth® is a registered trademark of Bluetooth SIG, Inc.

Lawrenceville, GA, USA   Chi N. Thai

Contents

Chapter 1: Introduction

1.1 ROBOTIS System Design Approach

For elementary and middle school students, ROBOTIS carries quite an extensive list of educational robotics kits for its internal market (http://www.robotis.com/index/product.php?cate_code=101010), but the USA and International stores carry a more limited offering (http://www.robotis.us/robotis-kits/), essentially the PLAY and DREAM systems. The PLAY system uses the sturdier 7 mm DIA rivet system which is better suited to motor skill levels of younger children. The DREAM system uses a smaller 3.5 mm DIA rivet system which will require more manual skills from the students (see Fig. 1.1).

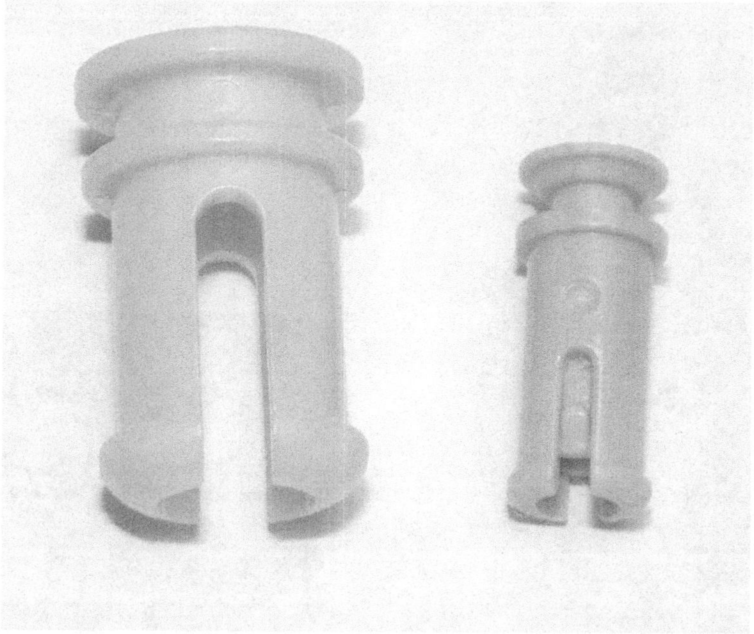

Fig. 1.1 ROBOTIS' rivet systems: 7 mm (left) and 3.5 mm (right).

The PLAY system (http://www.robotis.us/play/) is available in the USA in 3 versions: PLAY300, PLAY600 and PLAY700. The PLAY300 kit uses the older 3.5 mm rivet system and thus is not recommended for the youngest students. The PLAY600 kit is motorized and is quite popular for teaching wheel-based and linkage-based mechanical motions. It has been endorsed by the Global Educator Institute (http://geiendorsed.com/products/robotis-play-600-pets/). From Amazon reviews, most 6/7-years old users could handle this kit without adult supervision. Unfortunately, the PLAY600 kit is not programmable. The PLAY700 kit is also motorized with two continuous-turn motors and is equipped with 3 short-range NIR sensors and a speaker (for musical tunes only). The PLAY700 kit is programmable via smartphones and tablets (iOS® or Android®), as well as via Windows® PCs. With the included BT-410 module, it is fully programmable from mobile devices running its R+m.TASK and R+m.PLAY700 Apps and as such can add mobile services such as graphics, audio, video and speech to its robotic repertoire. Additionally, if the user has access to a BT-210 module or a BT-410 USB dongle, then the PLAY700 kit can be used with a Windows PC and interfaced to work with MIT's SCRATCH® 2 software.

The DREAM system (http://www.robotis.us/dream/) is a big upgrade over the PLAY700 kit, as it offers a more extensive range of sensors such as Touch, Color, Magnetic, Passive IR and Temperature sensors, and most importantly it can use 2 additional servo motors (not available with the PLAY700 kit) for Position Control applications, for example a pan-tilt connection for pointing a NIR sensor in a wanted direction or a pincer device to grab objects. It also works with the R+m.TASK and R+m.PLAY700 Apps which allow it to additionally access multimedia services on mobile devices.

Although different in their hardware features, both PLAY700 and DREAM systems share a common System Design Paradigm which considers a typical ROBOTIS robot to be a Computer Network with four major components:

1) A Hardware Controller, CM-50 for PLAY700 and CM-150 for DREAM (see Fig. 1.2), which is the "Main Brain" for the robot. It contains the user's instructions for the robot to execute the tasks wanted by the user.

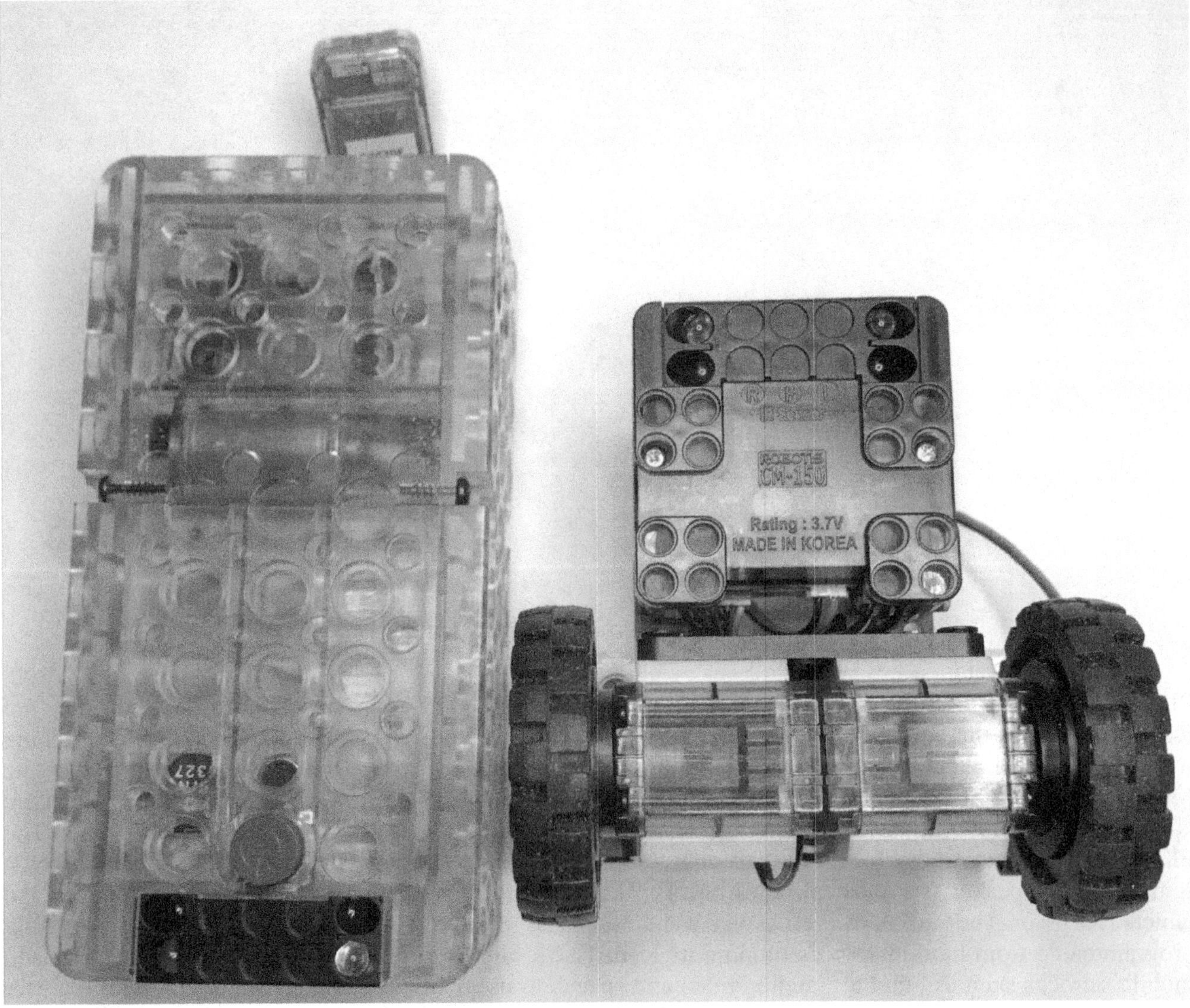

Fig. 1.2 Hardware Controllers: CM-50 (left) and CM-150 (right).

2) The Sensors component (e.g. NIR or Touch sensors – see Fig. 1.3), which helps the Hardware Controller get information about its surroundings. The more sophisticated sensors have their own mini-brains to communicate with the Hardware Controller.

Fig. 1.3 Selected Sensors: NIR Sensor (left) and Touch Sensor (right).

3) The Actuators component (e.g. Servo Motors or LEDs – see Fig. 1.4), which allows the Hardware Controller to perform the appropriate robot actions upon the real world as programmed by the user. Similarly, advanced actuators such as servo motors have their own mini-controllers to work in concert with the Hardware Controller.

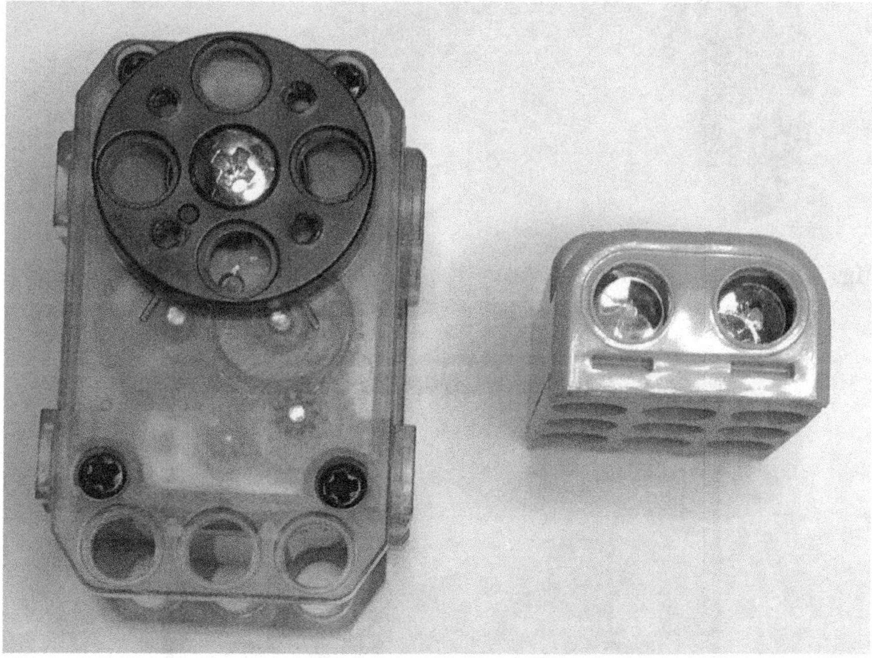

Fig. 1.4 Selected Actuators: Servo Motor (left) and LED Display (right)

4) A Remote Device, which can be an RC-100 module or a smartphone (see Fig. 1.5) or even a Windows PC. The Remote Device has a more flexible role depending on the specific device used and on its current role in this network of robotic components. For example, if the RC-100 is used with the DREAM system, it essentially functions as a wireless Sensor (attached to the user) informing the Hardware Controller about which "Up/Down/Left/Right" buttons had been pressed by the user. When a smartphone (running the R+m.PLAY700 App) is used with the CM-50, it can act as a Sensor telling the

CM-50 about which Touch Area on the phone screen had been pressed by the user, or it can act as an Actuator displaying appropriate graphics or texts as commanded by the CM-50 according to its programmed logic. When a PC is used as a Remote Device, the issues become more complex. For example, if TASK's Virtual Controller (see Fig. 1.6) is used, then the PC acts as a glorified Sensor informing the Hardware Controller CM-50/150 about which keyboard buttons had been pushed by the user. However, if the SCRATCH 2/R+SCRATCH tool chain is instead used on the PC, then the PC takes on the role of a "true" Hardware Controller, as SCRATCH 2 codes run on the PC and not on the CM-50 controller which now has a diminished role, i.e. just letting SCRATCH 2 commands from the PC to pass through to Actuators and Sensors via the CM-50's Firmware (see Fig. 1.8).

Fig. 1.5 Selected Remote Devices: RC-100 (Left) and Smartphone (right).

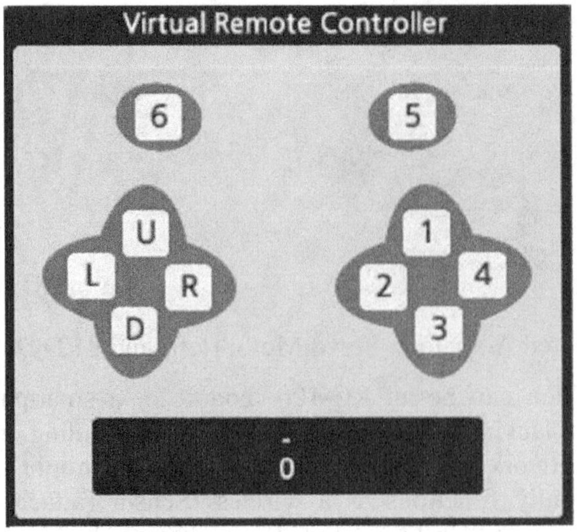

Fig. 1.6 TASK's Virtual Controller Interface.

To achieve hierarchical control in this adaptive network of robotic components, **ROBOTIS software uses two types of communication packets: Instruction and Status**. Instruction packets can only be issued by the Top-Level Controller which can be either the Hardware Controller (CM-50/150) or the Remote Device, and there can be only ONE Top-Level Controller at any one time (if ones use the proprietary firmware provided by ROBOTIS for its controllers). Once a network component receives an Instruction packet, it will execute the prescribed instruction and will optionally send back Status packets to the originator of that Instruction packet.

Fig. 1.7 shows a configuration of network components and communication packets flow for the situation where the Top-Level Controller is the Hardware Controller (CM-50/150) using the RoboPlus software suites (i.e. MANAGER, TASK, MOTION), while the Remote Device could be any of the following devices: physical RC-100, virtual RC-100 on the PC, smartphone or tablet using the R+m.PLAY700 App. In this configuration, please note that only the Hardware Controller can issue Instruction packets and that Actuators and Sensors can never issue Instruction packets as they are at the "bottom" of the control hierarchy.

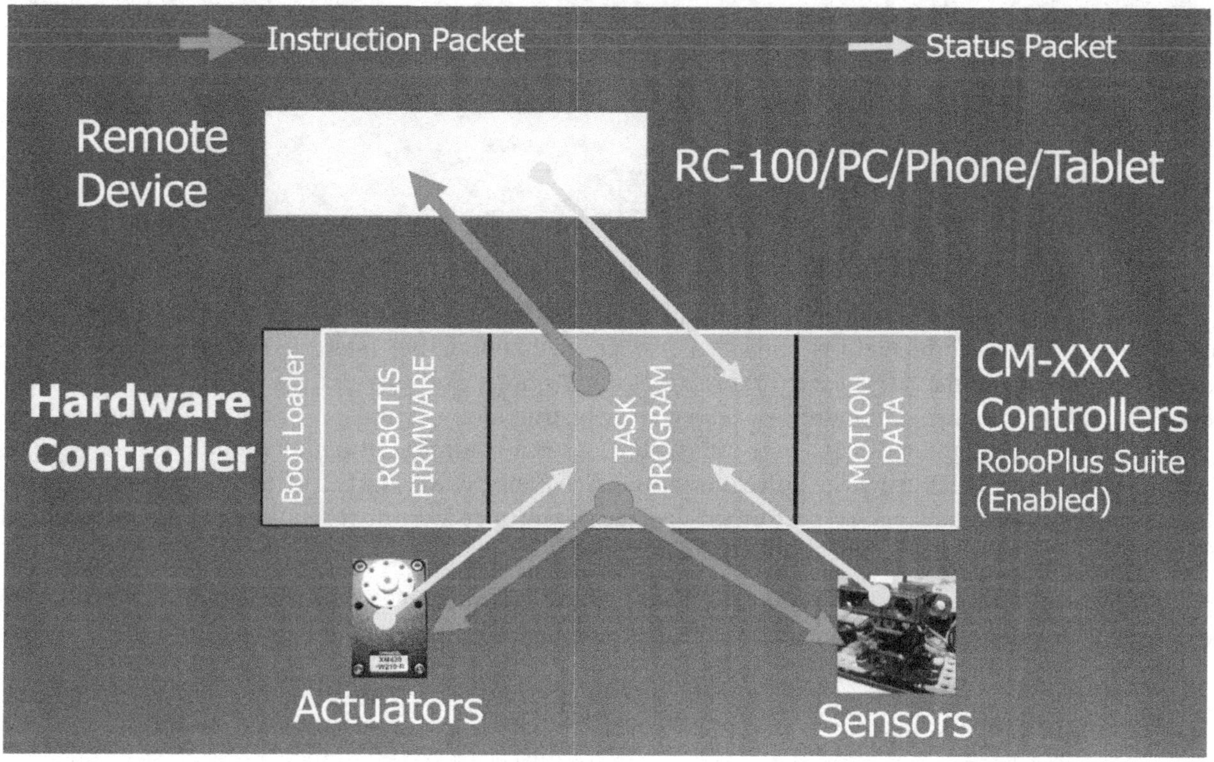

Fig. 1.7 Hardware Controller as Top-Level Controller.

Fig. 1.8 shows a second network configuration where the Top-Level Controller is the Remote Device which can be a PC using the SCRATCH 2/R+SCRATCH tool chain or a mobile device using a more advanced ROBOTIS software tool called OLLOBOT SDK (which is beyond the scope of this book and interested readers are referred to Thai (2017) for more details). In this configuration, please note that all TASK and MOTION functionalities would now be disabled on the CM-50/150 (mainly because ROBOTIS controllers' firmware are written this way). Please also note that all communication packets are now routed through the R+SCRATCH application on the PC and the "Firmware" component of the controller before reaching the Actuators and Sensors, meaning that communication delays would likely occur for this configuration due to the extra layer of communication software (especially when serial communication through R+SCRATCH is restricted to a rate of 57.6 Kbps).

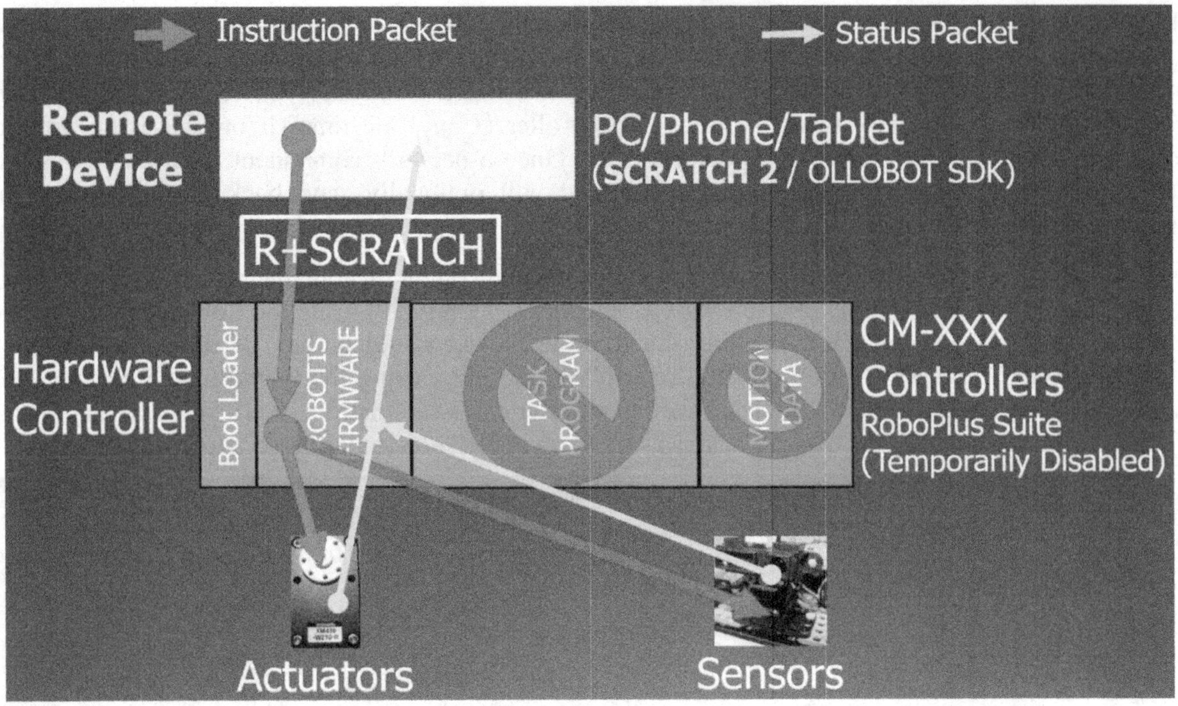

Fig. 1.8 Remote Device as Top-Level Controller.

There is a third configuration for a more advanced and flexible use of ROBOTIS robotic systems with Embedded C, which unfortunately is outside the scope of this book but interested readers can refer to this web link for more details (https://www.cntrobotics.com/robotis-evolving-dynamixel-concept). Advanced readers can learn more about Instruction and Status packets at (http://support.robotis.com/en/product/actuator/dynamixel_pro/communication/instruction_status_packet.htm). In this book, we'll be using the software tools TASK and SCRATCH 2 wherein the task of handling Instruction and Status packets is performed by the Controller's Firmware **"behind the scene"**, thus fortunately readers won't have to deal with them directly.

Please note that there is a third type of packet, called "RemoCon", used in Remote Control schemes whereas the user will have the opportunity to have a more hands-on practice in Section 3.6.

1.2 Sense-Think-Act Paradigm

Usually the younger robotics students were surprised when I started my robotics short courses by mentioning that they had been using the Sense-Think-Act paradigm in everyday activities by themselves. I used Fig. 1.9 to explain how humans use our Senses to let us know about the external World (i.e. Perception) and apply those Sensations into our Cognitive process (i.e. Thinking) to devise proper Actions/Reactions to the situation at hand. These Actions/Reactions in turn would change some aspects of the external World resulting in new Sensations triggering the next World-Human-Interactions cycle and so forth.

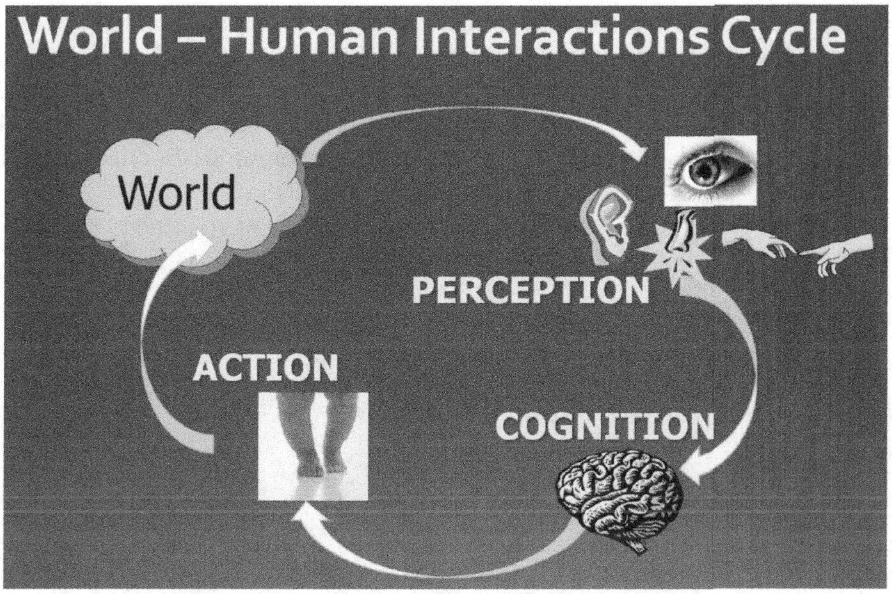

Fig. 1.9 World-Human Interactions Cycle.

Thus, in a way, doing Robotics is just trying to reproduce this interactions cycle onto a robot as shown in Fig. 1.10. Mechanical or electronic Sensors would convert their interactions with the World into Input Data to be sent to the Robot/Computer which was previously Programmed by the user to deal with Events deemed important to the operation of the Robot. Using this programmed Logic, the Robot Outputs appropriate commands to its Actuators which then could change the Sensors' perspective onto the World and therefore generate new Input Data for the Robot/Computer to process anew, through another interactions cycle.

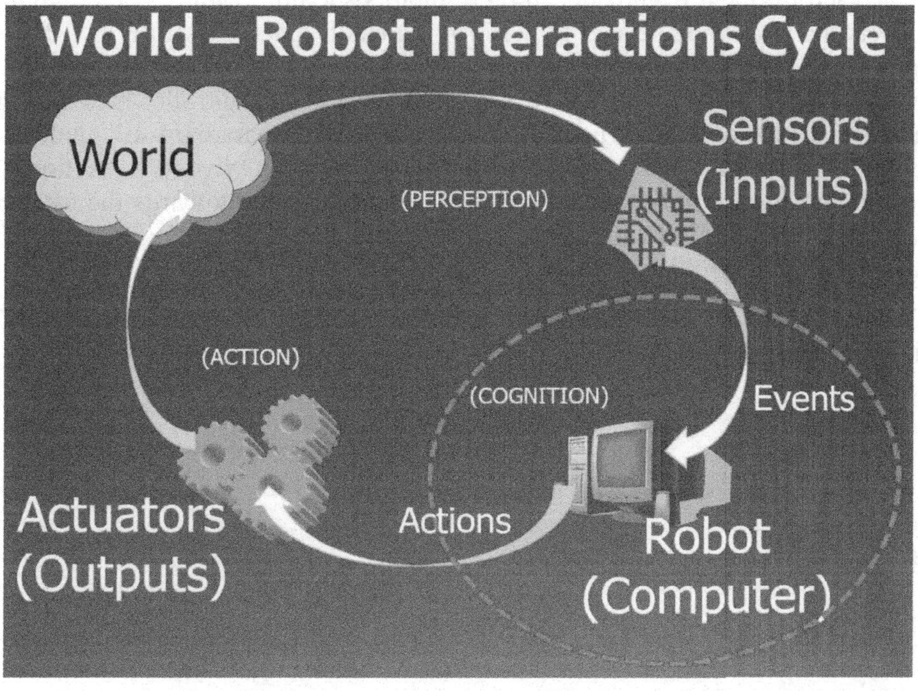

Fig. 1.10 World-Robot Interactions Cycle.

Let's go through an example to see how this Sense-Think-Act paradigm can help us solve the following challenge as described in Fig. 1.11, where a CM-50 based CarBot would be able to follow an irregular black track in the shape of the number 8. Furthermore, if the CarBot encounters a frontal obstacle during its travel, it would need to avoid it by swinging around until it gets back onto the track but in the other direction. Next, the CarBot should continue its auto-tracking mission unless it encounters a frontal obstacle again.

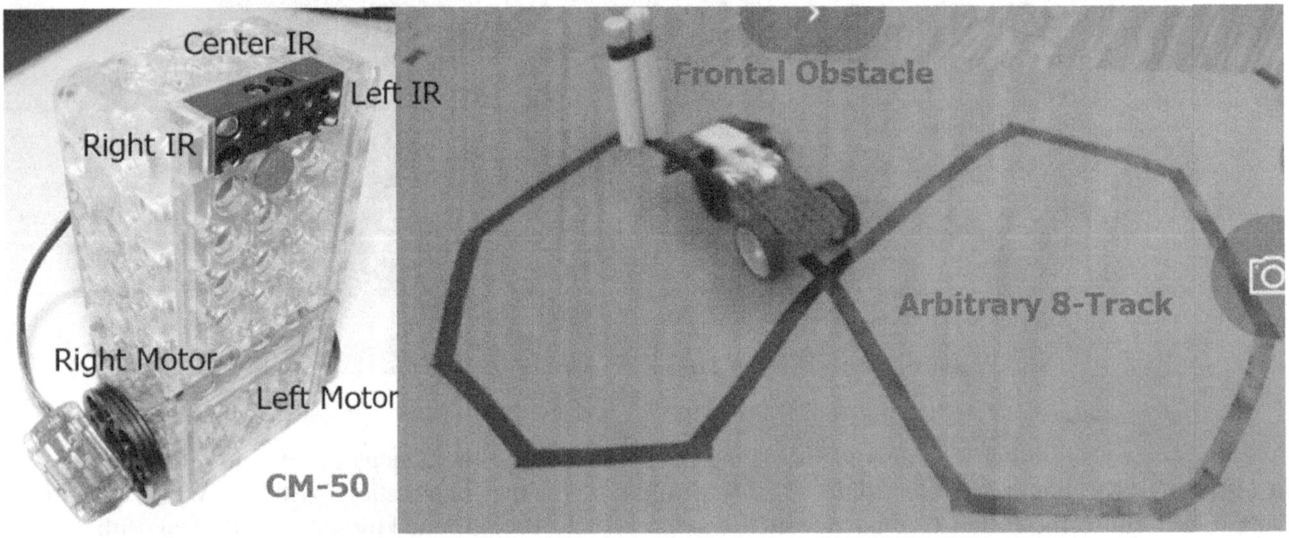

Fig. 1.11 CM-50 CarBot following an 8-track and avoiding frontal obstacle.

Fig. 1.11 shows that the CM-50 has three IR Sensors whereas the Center IR could be used to detect a frontal obstacle and the Right and Left IR sensors could be used to detect the curvy black track. The CM-50 also has a Motor on each Left and Right side, so these motors should be able to be programmed to provide all CarBot maneuvers that would be needed to perform the challenge successfully (i.e. Forward, Turn Left and Turn Right). This System Design solution seems feasible enough, but to progress further towards the actual coding of the CM-50, we need to zoom in around the Robot/Controller area in Fig. 1.10 (i.e. the dashed red boundary in the lower right corner). The Sense-Think-Act paradigm still applies at this level, but some robotics practitioners would start using different labels such as "Events", "Actions" and "Reactive Control" as they are more useful in the next steps that we need to take towards the creation of the actual computer code to control this CarBot. Interested readers should read Matarić (2007) for a more complete description of other robotics control approaches.

With the "Reactive Control" approach, ones need to match up a possible Condition/Event with an appropriate Action for the robot to do (see Fig. 1.12).

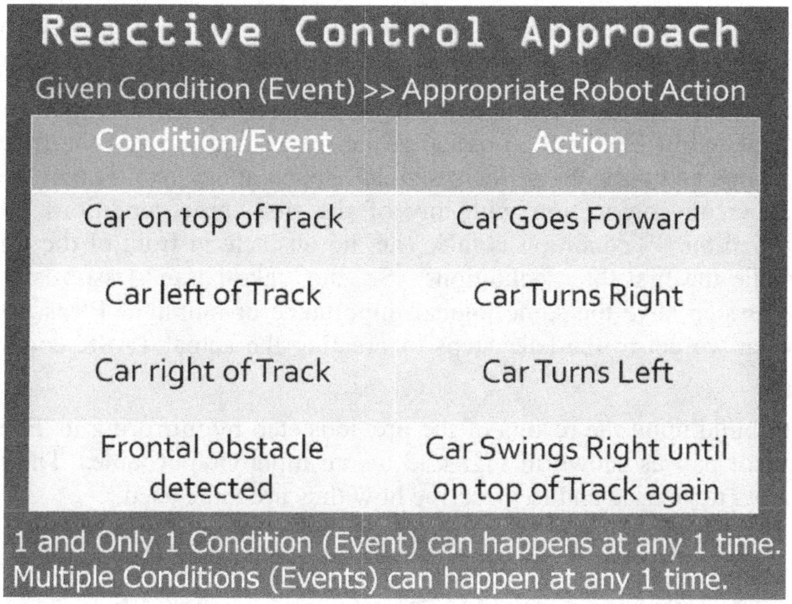

Fig. 1.12 Reactive Control Approach used for CarBot challenge.

For example, if the CarBot happens to be lined up on top of the track already, then it should just go forward. There are 3 other Event/Action pairs listed in Fig. 1.12, and are these four pairs the only ones to consider or needed for this robot challenge? My advice for beginning roboteers is just to start with a reasonable set of Event/Action pairs and then improve on them as needed in later refinement steps.

But how would a beginner in robotics comes up with such a list as shown in Fig. 1.12 in the first place? Here, in keeping with the introductory scope of this book, the author offers a simple "thought experiment" or "visualization exercise" approach:

1) First, we would imagine to be "one" with the robot, so our Left and Right Eyes would match with the Left and Right IR Sensors, i.e. we keep our nose to the ground so that we can see the black track (to be exact, only a partial view of it). However, humans do not have a third eye right on top of our head to correspond to the Center IR, so we would really have to use our imagination and use our "mind's eye" as the third eye there!

2) So now that we are the robot, how would we go about performing this 8-track challenge? Here, although not mentioned in the challenge write-up, we need to decide on where is the starting position of the robot with respect to the 8-track? At the start of the challenge, the robot could very well be positioned on a brown paper area inside or outside the track, or it could straddle the track itself. As the challenge's rules did not specify the starting event, we are going to use the "easiest" option of having the CarBot start its challenge already lined up with the track. Then, the reader can see why the first Event/Action pair came about on the list in Fig. 1.12. Next, as the 8-track would curve here and there, the robot may find itself left or right of the track (second and third Event/Action pair respectively). So, with the first three Event/Action pairs, we should feel confident, at least logic wise, that the CarBot would be able to follow the 8-track using computer codes derived from Fig. 1.12.

3) Similarly, we can come up with the 4th Event/Action pair to take care of the situation when an obstacle comes within detection range of the Center IR. Here the alert reader would notice that the first three Actions were Simple Actions triggered once certain conditions occurred. But the 4th one was an

"Action-until-Event", i.e. Act now until some Event happens later. In other words, there is sort of a nested Event/Action inside an "Action-until-Event".

Before we go on to the next step, we need to have a closer look at the four Event/Action pairs (or Conditions if you will) listed in Fig. 1.12. Logically, the reader can see that the first three conditions are mutually exclusive (i.e. one and only one of them should happen at any one time in "real life"), but that the 4th condition may happen in conjunction with any of the first three conditions. Or using a different viewpoint, it is only when the 4th condition is false (i.e. no obstacle in front of the CarBot) that we would care about how to handle the first three conditions. So, the "take-home" lesson is that not all conditions "discovered" during this step have the same logical importance or ranking. Please keep this hierarchy of conditions in mind when we get to the later steps of creating the actual TASK code for this challenge in Chapter 3.

The next step is to build upon the results of the previous step by mirroring its Event/Action pair into a matching Sensor/Actuator pair as shown in Fig. 1.13, as an Input/Output table. The goal is to specify the kind of Sensors/Actuators to be used and to prescribe how they are to be used.

Input-Output Table
Given Input Sensor(s) >> Activate Appro. Actuator(s)

Input Sensors	Output Actuators
Right IR & Left IR see "Black"	Right Motor CW Left Motor CCW
Right IR sees "Black" & Left IR sees "Brown"	Right Motor Stops Left Motor CCW
Right IR sees "Brown" & Left IR sees "Black"	Right Motor CW Left Motor Stops
Center IR sees "Obstacle"	Right Motor Stops Left Motor CCW until Left IR sees "Black"?

Fig. 1.13 Input/Output Table mirroring list of Event/Action pairs as shown in Fig. 1.12.

For example, let's look at the second Sensor/Actuator pair in Fig. 1.13 which corresponds to the second Event/Action pair in Fig. 1.12 which specifies that if the CarBot is found on the left of the black track then it should turn right so that it can get back onto the track. Thus, for the CarBot to be on the left of the track, its Left IR should be seeing "Brown" and its Right IR should be seeing "Black" (see 2nd row of the Input Sensors column in Fig. 1.13). The corresponding Action for the CarBot is then to turn right, i.e. the Right motor should stop and the Left motor should keep on going "Forward", i.e. keep on turning CCW (see 2nd row of the Output Actuators column in Fig. 1.13).

As an exercise to the reader (a pop quiz, if you will), let's have a closer look at the specifications for the Output Actuators for the 4th Event when a frontal obstacle comes within detection range of the Center IR. Here, the CarBot's Action is supposed to swing right until it straddles the black track again but in the reverse direction, so the specifications were to stop the Right motor and to keep the Left motor going

"Forward" (i.e. CCW) so that the CarBot would turn right, until the Left IR sees "Black". Or would the reader think that it should really be "until the Right IR sees Black" instead? The answer will be given in Chapter 3.

Please note that the previous procedures do not involve the use of any computer nor learning about how to use a specific software tool, and the author highly recommends that beginner roboteers to always start with these procedures instead of jumping directly into coding or try to adapt someone else codes, as these procedures would help the roboteer refine the tasks to be done by the robot before committing them into hard codes, thus saving the programmer lots of time in the long run.

Another benefit of using the Event/Action and Input/Output tables is that they will help with the actual writing of the TASK code (as shown later in Chapter 3) as the ROBOTIS syntax used would support practically a one-to-one correspondence of all Event/Action pairs A to Z from these tables into similar "physical" locations within the final program, except for two aspects that had not been discussed before (see Fig. 1.14):

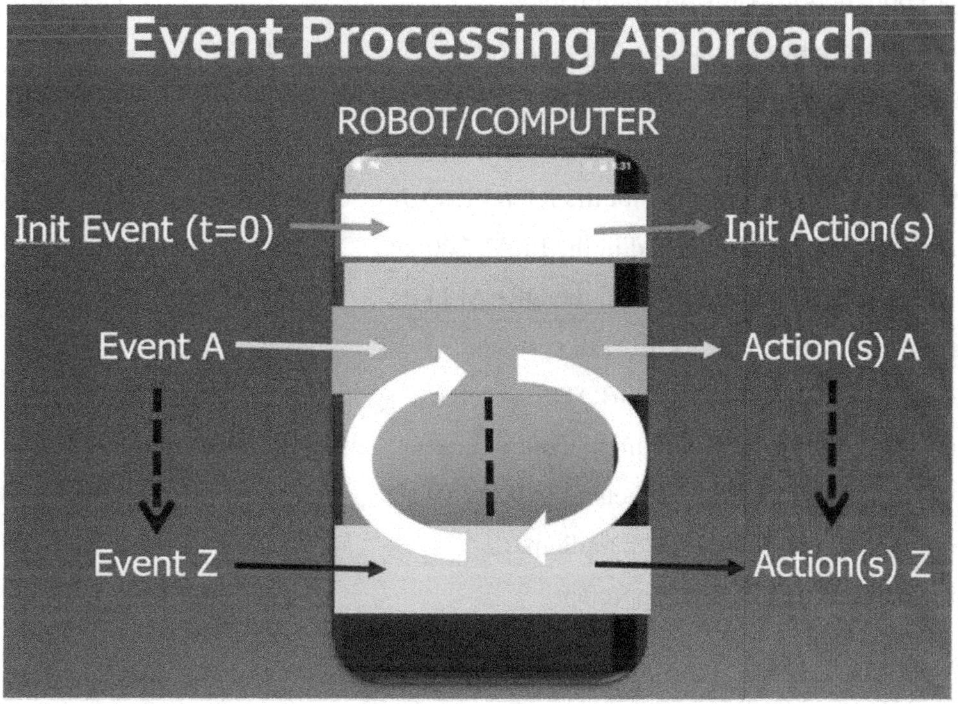

Fig. 1.14 Event/Action pairs matching locations in TASK code.

1) A main code looping structure that allows the controller to go through the list of possible Events A to Z that may occur during run time, as it listens to what the Sensors tell it about the current state of the World. This is the overall loop shown in Figs. 1.9 and 1.10.

2) The Init Event which occurs once at time=0, i.e. when power is applied to the robot or when the TASK program starts running. The Init Event would usually result in a known state for the robot which could be some audio output to let the user knows that it is ready for "action", or it could be a display of a menu of tasks to let the user choose which task for the robot to perform.

In the next Chapter 2, we'll get into the details of setting up the hardware and software for the PLAY700 kit and show the readers how to use various software tools on mobile devices and Windows PCs. The readers would also be shown how to use TASK and R+m.PLAY700 to run the demo programs that came with the PLAY700 Kit when using a mobile device.

1.3 References

Matarić MJ (2007) The robotics primer. The MIT Press, Cambridge.

Thai CN (2017) Exploring Robotics with ROBOTIS Systems (2nd Edition). Springer Publishers International, New York.

1.4 Review Questions for Chapter 1

1. What are the sizes of the rivet systems used on the DREAM and PLAY robotics systems?
2. The PLAY600 robotics kit is programmable. T F
3. The PLAY700 robotics kit includes servo motors T F
4. How many NIR Sensors can be used on the PLAY700 kit?
5. What is the name of the mobile software App that works with the PLAY700 system?
6. Does the DREAM system work with the R+m.PLAY700 software tool?
7. Name the hardware controller used in the PLAY700 kit.
8. Name the hardware controller used in the DREAM kit.
9. List examples of Sensors for ROBOTIS systems.
10. List examples of Actuators for ROBOTIS systems.
11. List items that can be used as Remote Controllers for the PLAY700 system.
12. List types of communication packets used in ROBOTIS systems.
13. Describe the "Sense-Think-Act" paradigm.
14. Describe the Reactive Control approach.

Chapter 2: Getting Started

The PLAY700 kit, as received, is fully functional with a mobile device on iOS® or Android® OS. This web site (http://www.robotis.us/software/play700/) contains download links for all needed PLAY700 software and tutorial videos. Ones can use the mobile R+m.TASK and R+m.PLAY700 Apps using Bluetooth 4 on the mobile device and a corresponding BT-410 module on the PLAY700 robot. Some features of the R+m.PLAY700 tool are only functional on an Android device, and not on an iOS device (see Section 2.1).

If the user wants to use MIT's SCRATCH 2 software with PLAY700, then the user is restricted to Windows PCs, as the needed ROBOTIS helper software named R+SCRATCH is available for Windows OS only. The user also needs to install the SCRATCH 2 Offline Editor (https://scratch.mit.edu/download) on the PC. The user is further required to purchase a USB BT-410 dongle from ROBOTIS (or their vendors) so that the Windows PC can communicate with the BT-410 on the robot (see Fig. 2.1).

Another handy ROBOTIS software tool called MANAGER also runs on Windows PCs only. It provides a quick way to check if the CM-50 is working OK hardware wise, and to update its firmware when necessary.

In this Chapter, we'll use a BasicBot as the demonstration platform (see Fig. 2.1). Please watch Video 2.1 to see how it is put together.

Fig. 2.1 BasicBot demonstration platform.

2.1 Using PLAY700 kit on Mobile Devices

The first step is for the reader to install the R+m.TASK and R+m.PLAY700 Apps on his or her chosen mobile device. These Apps are available at the appropriate App Stores (https://itunes.apple.com/us/developer/robotis-co-ltd/id948481761) or (https://play.google.com/store/apps/developer?id=ROBOTIS&hl=en).

Next, the reader is invited to view Video 2.2 which is a tutorial on how to:

1) Connect the mobile device to the BT-410 on a BasicBot platform. It also demonstrates other setup settings (see Fig. 2.2).

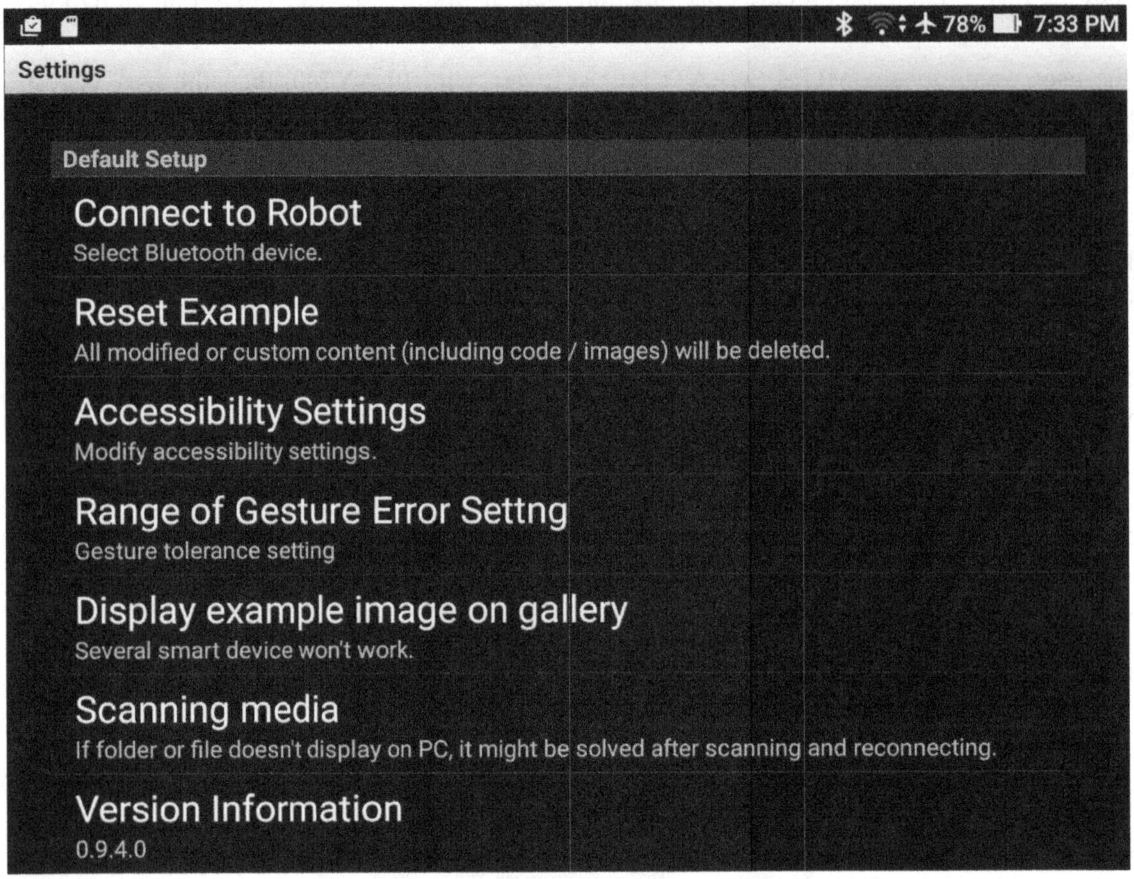

Fig. 2.2 R+m.PLAY700 App's Settings.

2) Explore the features available in a typical PLAY700 project and see the differences in capabilities between Android and iOS devices (see Fig. 2.3).

Fig. 2.3 Features available in a typical PLAY700 project.

3) Use the TASK tool to have a quick browse at the preloaded PLAY700 demo program, and show how to download it to the CM-50 and run it in concert with one of the 4 R+m.PLAY700 example projects on the mobile device (see Fig. 2.4).

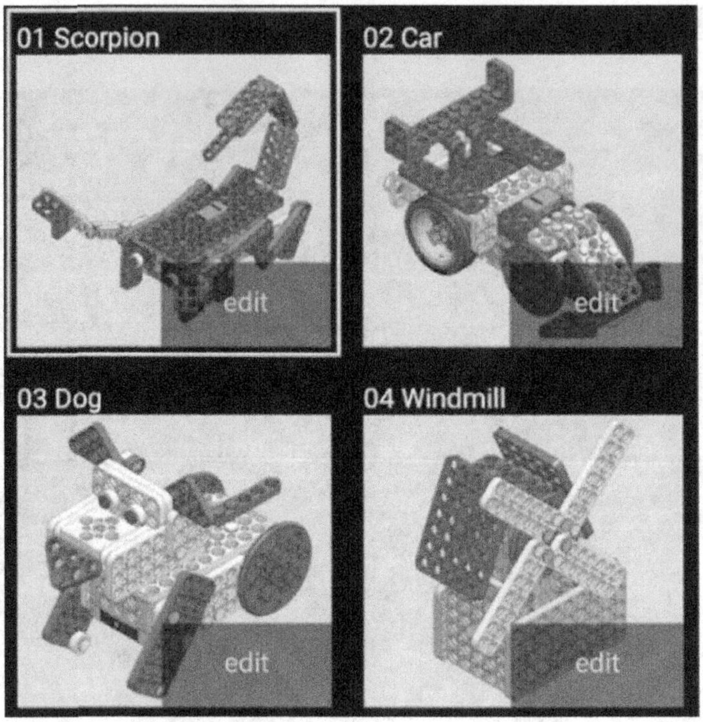

Fig. 2.4 R+m.PLAY700 App's Example Projects.

2.2 Using PLAY700 kit on Windows PC

Assuming the reader had installed successfully the following software packages on the PC:

1) MANAGER (http://www.robotis.us/roboplus2/).
2) TASK (http://www.robotis.us/roboplus2/).
3) R+SCRATCH (http://www.robotis.us/roboplus2/).
4) MIT SCRATCH 2 (https://scratch.mit.edu/download).

The MANAGER tool is used when the user needs to check if a given CM Controller and the actuators or sensors attached to it are working properly. Fig. 2.5 shows the main menu for the MANAGER tool when it connects to a CM-50 (i.e. the controller for the PLAY700 system). The column labeled "Control Table" contains various functions that can be tested directly via MANAGER: for example, AUX LEDs, Ports 1 & 2 Motor Speeds, etc... These functions can also be programmed via the TASK tool as shown later. Sometimes, the CM-50 may malfunction in some way, and usually the best solution is to "recover its firmware" which can only be done via this MANAGER tool. Please see Video 2.3 for more details.

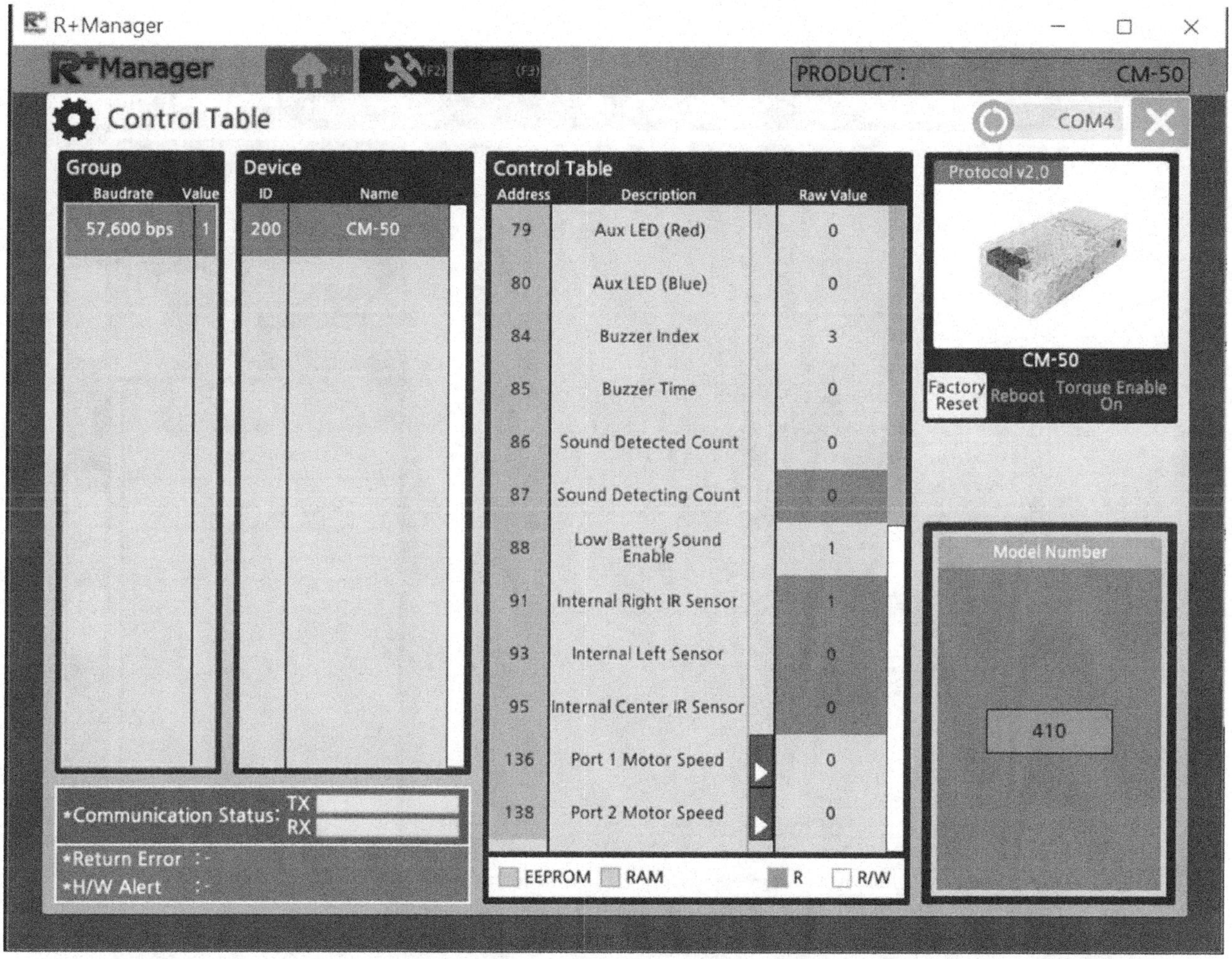

Fig. 2.5 MANAGER Tool on CM-50.

The TASK tool is THE "programming" tool, where the results from Input/Output tables, created from steps described in Chapter 1, are converted into TASK program statements. The TASK tool may be described as a context-sensitive line editor (see Fig. 2.6).

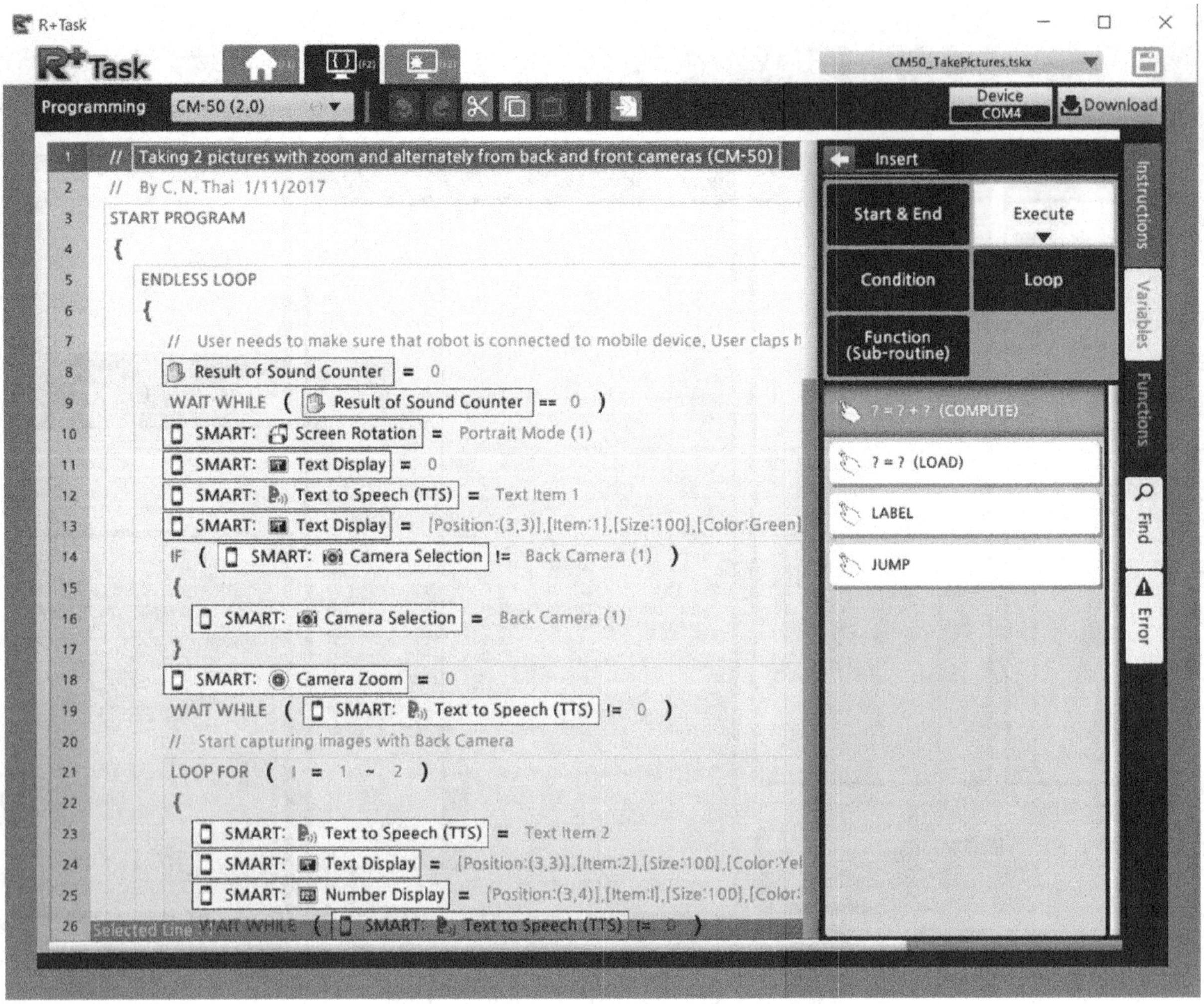

Fig. 2.6 TASK tool on a computer program for CM-50.

It has special "Smart Device" functions that interface with the R+m.PLAY700 App allowing the CM-50 to control various services on the mobile device such as graphics, audio, video, text-to-speech, speed recognition and image processing (see Fig. 2.7). Please see Video 2.4 for a quick tour of its capabilities. Its proficient use requires lots of training and practice on the part of the user which will be developed in Chapter 3.

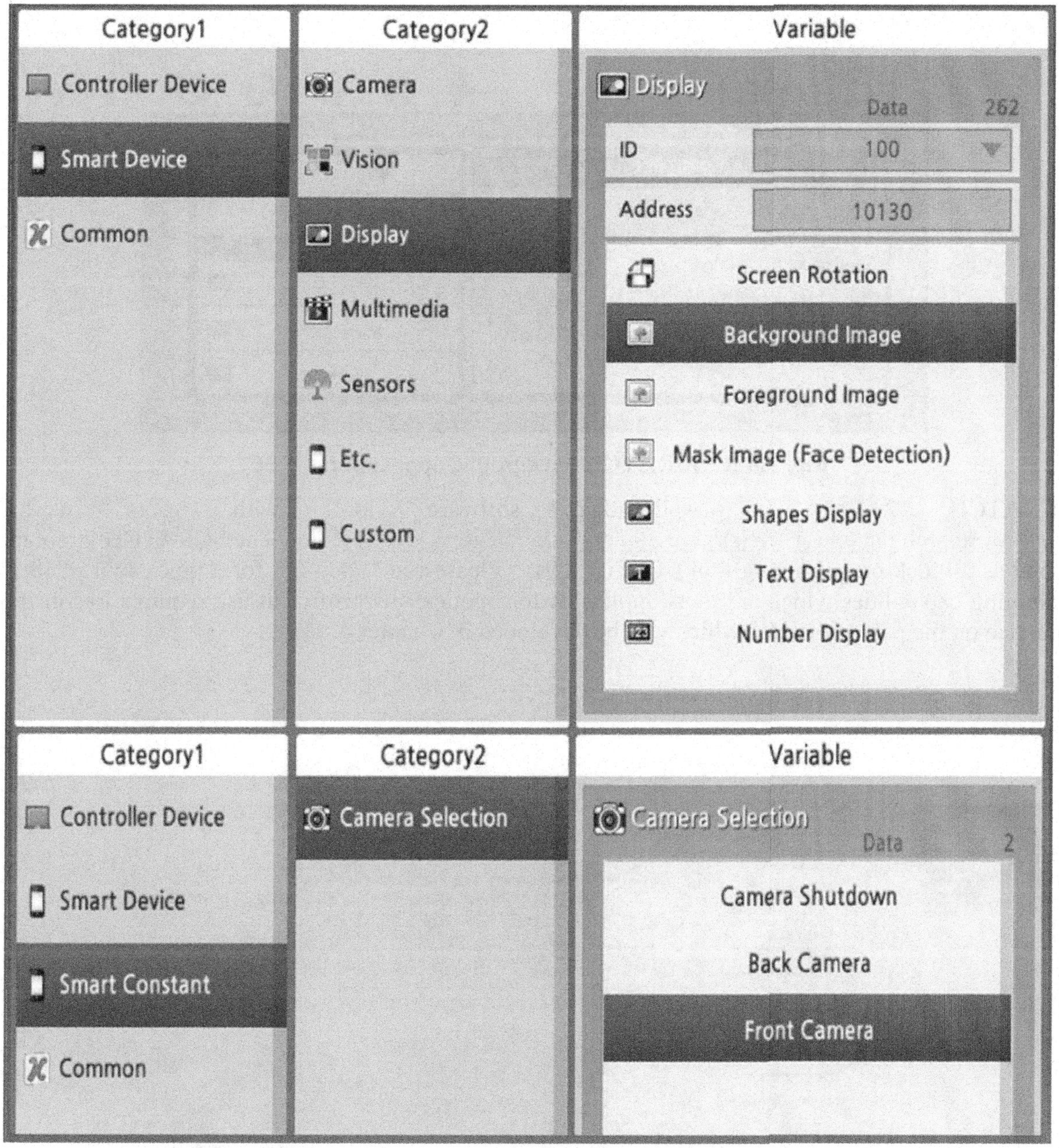

Fig. 2.7 SMART DEVICE functions available inside TASK tool.

R+SCRATCH (see Fig. 2.8) is an important background application that interfaces with the SCRATCH 2 software tool, to allow SCRATCH 2 commands to be sent from a PC communication port to reach into the actuators and sensors on the CM-50. Please watch Video 2.5 for a demonstration of its typical use.

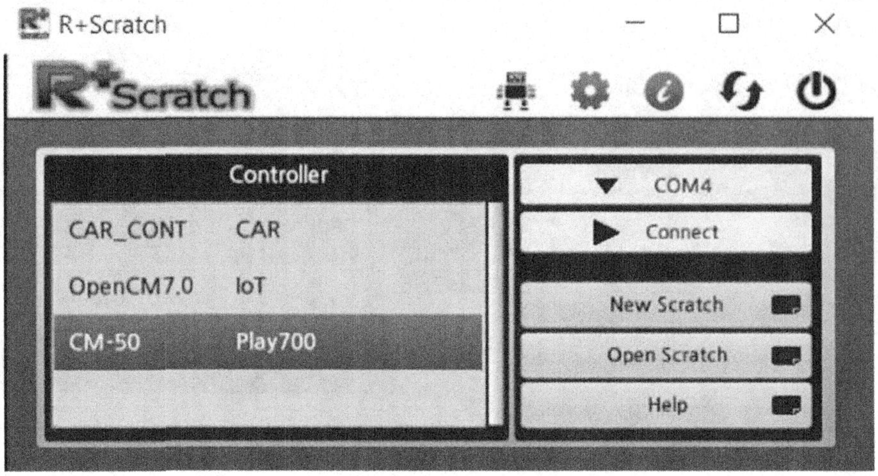

Fig. 2.8 R+SCRATCH background application for PC.

SCRATCH 2 is a well-known software tool with users world-wide (https://wiki.scratch.mit.edu/wiki/Scratch_Statistics#Scratchers_Worldwide). The SCRATCH 2 tool may be described as a block-programming tool (see Fig. 2.9). Please see Video 2.6 for a quick tour of its event-programming capabilities which are most applicable to robotics. Its proficient use requires lots of training and practice on the part of the user which will be developed in Chapter 4.

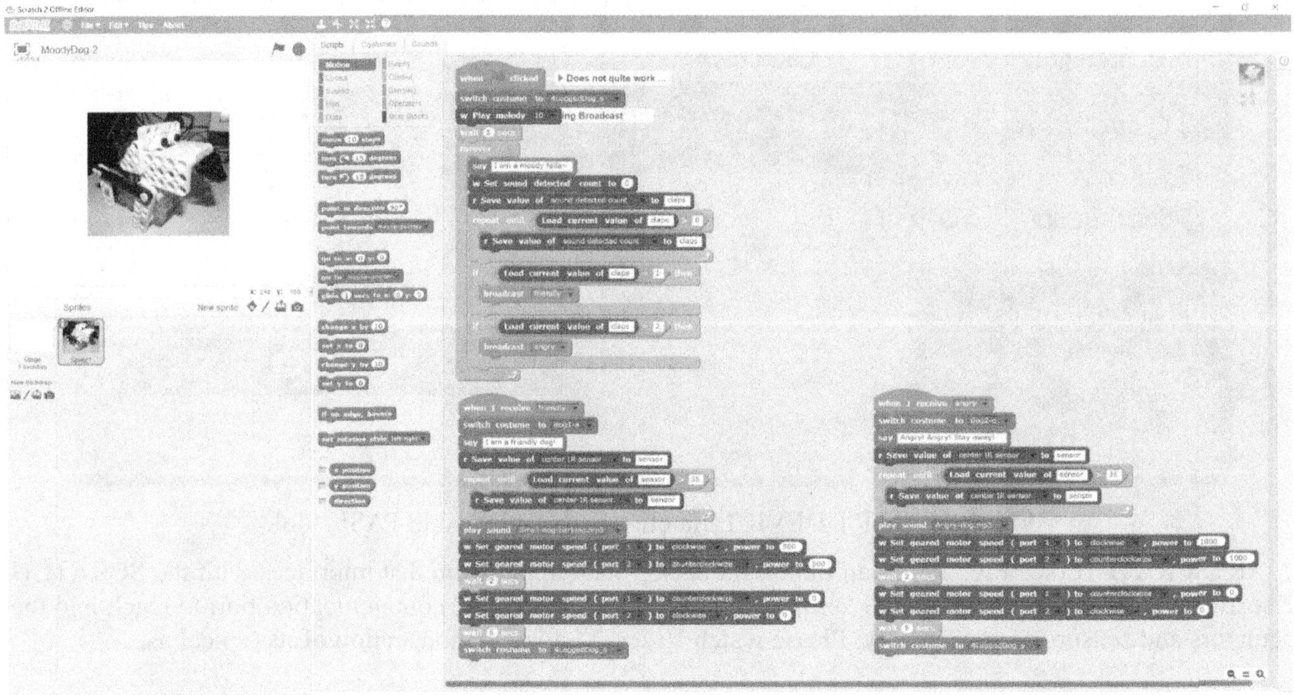

Fig. 2.9 SCRATCH 2 for Moody DogBot project.

2.3 Building "Spinning Top" robot

Let's build "Spinning Top" which will serve as the standard platform from where we can develop foundational programming skills using the TASK, R+m.PLAY700 and SCRATCH 2 tools (see Figs. 2.10 and 2.11, and watch Video 2.7 to see how "Spinning Top" was constructed from "BasicBot"). We'll use this robot to learn about robotics interfacing and control techniques in Chapters 3 and 4.

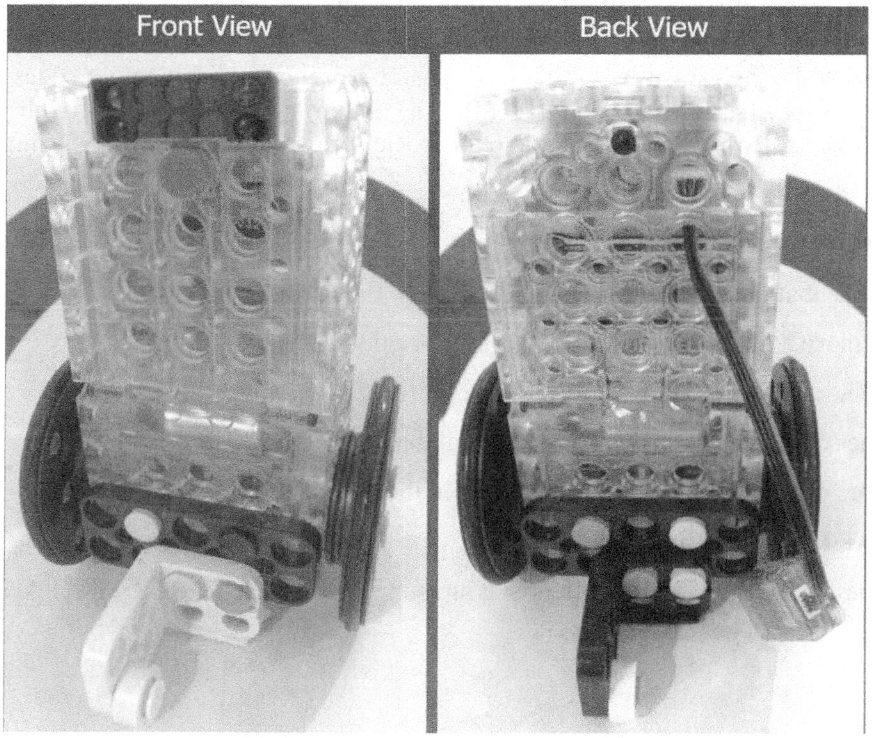

Fig. 2.10 Front and Back Views of "Spinning Top" robot.

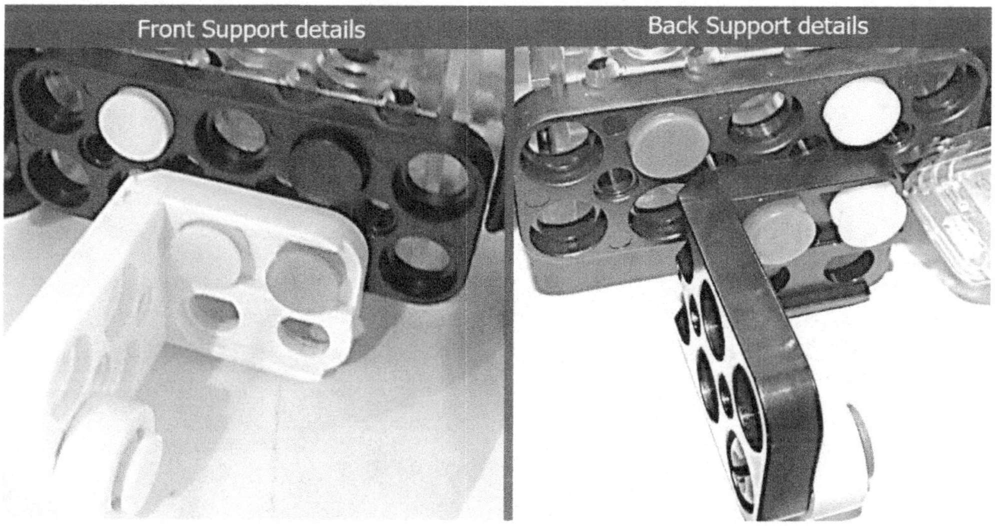

Fig. 2.11 Details of front and back support brackets.

2.4 Downloading Example Codes

For PCs and Android devices, the reader needs to first visit this DropBox link (https://www.dropbox.com/s/mecq2yuvsrfixfr/LearningRoboticsWithRobotisPlaySystems_Codes.zip?dl=0) and download the ZIP file to the Windows PC, next unzip it and transfer all the example codes from the PC to the Android device via Windows Explorer (please see Part 1 of Video 2.8). For iOS devices, the procedure is more involved as the Windows PC normally can only "see" the Photo Gallery on these devices. The author uses the "iMazing" software which can transfer seamlessly the example codes from the PC over to the iPhone file directory (see Part 2 of Video 2.8). The author has not done a thorough Web search to see if there are some "free" software/app that can do similar files transfer tasks as with iMazing.

2.5 Review Questions for Chapter 2

1. List 5 features that are available for a project inside the R+m.PLAY700 App.
2. List the memory address for the Center IR Sensor on the CM-50 Controller.
3. List the memory address for the Port 1 Motor on the CM-50 Controller.
4. List the types of "EXECUTE" commands available in the TASK tool.
5. List the types of "LOOP" structures available in the TASK tool.
6. List 3 SMART DEVICE functions.
7. List the types of ROBOTIS Controllers that are supported by the R+SCRATCH App.

Chapter 3: Using R+TASK & R+m.PLAY700

This chapter is written for a complete beginner in robotics and computer programming, thus if the reader already had some experiences in those areas, please bear with the "pedestrian" approach used and feel free to skip to the more advanced sections using the R+m.PLAY700 mobile App as needed. We will be using the "Spinning Top" platform (Fig. 2.10 from Section 2.3) to learn how to use the TASK tool and the R+m.PLAY700 App. A "Spiral" learning approach (Bergin, 2012) will be used for instructional purposes where the author would first show how to use the TASK tool with the "plain" CM-50 to perform a certain project. Next, the author would "enhance" it with similar or enhanced features provided by the R+m.PLAY700 App (whenever appropriate).

The TASK tool offers most of the standard programming structures found in other languages such as C/C++. Below is a short description of the most useful features for a beginning programmer (more details on specific items will be provided at appropriate sections in this chapter):

1) ASSIGNMENTS – TASK provides two types of assignment statements: LOAD and COMPUTE.

The syntax for LOAD is "A = B" with the usual meaning of "Operand A is assigned the Value of Operand B".

The syntax for COMPUTE is "A = B *Op* C" where the RHS represents the Value obtained when performing the "*Op*" operation between Operands B and C. The "*Op*" operations supported are the 5 basic arithmetic operators "+, - , * , / , and % (remainder)", and the 2 bit-level operators such as "&" (AND) and "|" (OR). Only INTEGER arithmetic is supported. VARIABLE parameters are supported but arrays are not.

2) LABEL and JUMP statements are supported.

3) CONDITIONS – For Conditional statements, TASK provides the usual logical operators: "&&" and "||" respectively for logical AND and OR operations, and "==", "!=" , "<", "<=", ">", ">=" for equality and inequality tests. The standard IF, IF-ELSE-IF, ELSE structures are supported, but nested conditional statements using parentheses are not supported (so the user will have to apply DeMorgan's theorems to expand complex logical statements as needed).

4) LOOPS – for Repetition statements, TASK provides "WAIT WHILE", "LOOP WHILE", "LOOP FOR", "ENDLESS LOOP" and "BREAK LOOP".

5) FUNCTION definition and calling are provided in TASK. A special CALLBACK function can also be used: it is triggered by a hardware timer every 8 ms independently of the main TASK program.

3.1 Learning Sequence Control in TASK programs

Please review Video 2.4 if you had not done so to refresh your memory about navigating the menus of the TASK tool on a PC or mobile devices.

In this unit, we are going to learn that by default the CM-50 controller behaves sequentially, i.e. it can do only one thing at a time. Video 3.1 shows how to create a TASK program from scratch, please save this program as "SequenceControl-1.tskx". Video 3.1 also shows how this work was done on PCs and on mobile devices (Android and iOS). Alternately, the reader can use the TASK tool to open the same pre-written program from the provided ZIP file containing all the example codes for this book (see Section 2.4). Fig. 3.1 is a screen capture of this program which has only 3 lines of "relevant" code (lines 5-7). More formally, they are called "Statements". All TASK program begins with the "START PROGRAM" statement (line 3) and an opening curly bracket "{" statement (line 4), and ends with a closing curly bracket "}" statement (line 8).

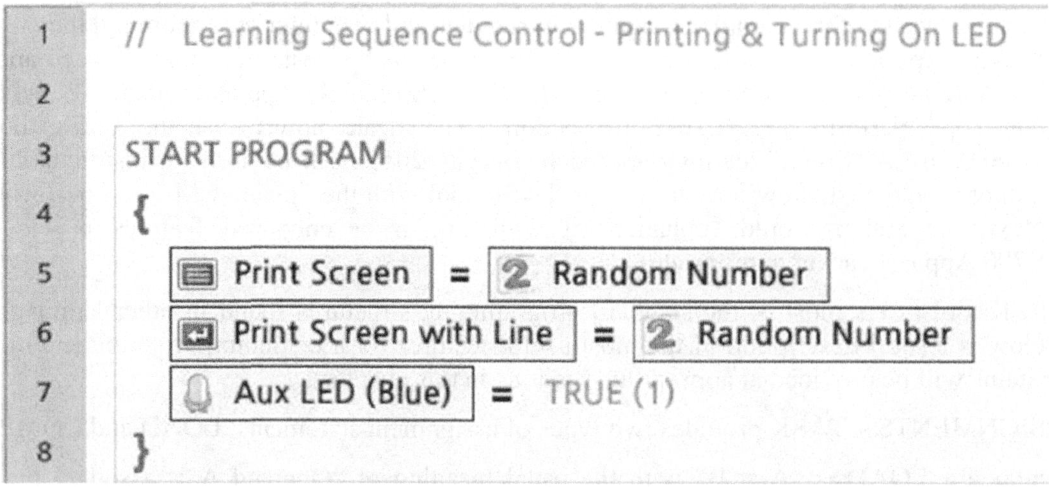

Fig. 3.1 Program "SequenceControl-1.tskx".

All statements between the two matching curly brackets "{}" are considered "legitimate" instructions (but may not be "logically correct") for the CM-50 controller to execute in sequence, top to bottom, at run time. But if they are preceded by the double forward slashes, "//", then they are considered as "Comments", i.e. information for the human reader only and not actual instructions for the controller. For example, Line 1 just tells the reader about the goals of this program. Later, the reader will be shown that it is sometimes necessary to "nest" the curly brackets to create compound statements to accomplish certain logical flows needed in a wanted algorithm.

Statements 5 to 7 are called LOAD statements which have this form (i.e. syntax): "A = B". The "=" sign is not the regular "equality" sign that the reader may have seen in algebra. In the world of computer programming, the "=" symbol stands for an "Assignment" operator. Each of the letters/symbols "A" and "B" is called "Operand". Thus "A=B" should be read as ("A" is assigned with the Value of "B"), or it can also be read as ("Value of B" is assigned to "A"). Essentially "B" should be a Data Value (e.g. TRUE or "1" – see Statement 7) or a Source of Data Input (e.g. a Random Number generated by the Controller internally – see Statements 5 and 6). In other words, the beginner programmer should get used to "read" a LOAD statement from the Right-Hand-Side (RHS) to the Left-Hand-Side (LHS) (which may be "unusual" to English readers). As a side note, TASK supports a second Assignment type of statement called COMPUTE (see its use in Section 3.2) and practically 90% of all the Statements in a typical TASK program are of the "Assignment" type.

When the controller executes Line 5 at run time, it will want to execute the RHS first, i.e. it will go to a specific memory location (Address 77 to be exact – and you may remember this from the previous Video 2.3) and it will pull out whatever random number between 0 and 255 that it can find there. Next, it will "assign" this found value to the LHS which is the Screen Printer. The user can then see this random number printed out on the Output Screen of the PC or mobile device, but the display cursor will stay on the same output line afterwards, because the wanted command is a "Print Screen". Line 6 will execute similarly to

Line 5: i.e. it will print a new random number on the same line on the output screen but then it will execute a "carriage return" to go to a new line afterwards, because the wanted command is now "Print Screen with (new) Line".

Statement 7 is also an Assignment statement where a "TRUE" value is assigned to the Blue LED: i.e., you guessed it, to turn it ON.

Please see Video 3.2 for more details on how this TASK program "SequenceControl-1.tskx" would actually behave at run time on a Windows PC, then on a mobile device.

Video 3.2 shows the user that because of all the time delays needed for Bluetooth connections to work, the run time behavior of this TASK program was far from satisfactory. On the PC, we saw the display of the random numbers, but we did not see the Blue LED turn itself on during run time. On mobile devices, we could not see any "Print Screen" output at all. The "programming logic" used was correct as the controller could compile and execute its machine code perfectly fine, so what was going on with our robot? **The key understanding here is that everything in real life, and therefore in robotics also, "takes time" to get done, so we'll have to figure out a way for the controller not to "rush through" its program, so that we, slow-poke humans, could "see" what is going on.** So, we'll need to figure out a way for the controller to do "nothing" for a user-defined time slot during which Bluetooth devices can establish their connections.

The "SequenceControl-2.tskx" program implements such as scheme using the High_Resolution Timer available on the CM-50 controller. Fig. 3.2 shows this completed program and Video 3.3 shows how it can be created by modifying the previous program "SequenceControl-1.tskx".

Lines 6 and 7 are new Assignment statements where "5 sec" is assigned to a Variable Parameter named "Overall_Time" and where "4 sec" is assigned to a second Variable named "BT_Connect_Time".

The Hi-Res Timer is introduced here and it is used twice (lines 8-9 and 11, and lines 15-16). It functions similarly to a stop-watch or a timer on a microwave oven. To use the Hi-Res Timer, we need first to load the Hi-Res Timer Counter with the wanted time delay (4 seconds in line 8 and line 15) – remember Address 74 from Video 2.3? The Hi-Res Timer Counter would then automatically start to count down to zero with a 1 ms decrement. Next, we need to use a WAIT WHILE statement such as in lines 9, 11 and 16 to create user-defined time slots associated with certain events. The WAIT WHILE statement is "unusual" in the way that it keeps on executing itself, i.e. "does nothing", as long as the CONDITION specified within the two parentheses "()" remains TRUE, for example (Hi-Res Timer Counter > BT_Connect_Time) in Line 9. Let's now do a "thought experiment" to see how Lines 8 through 11 can work together at run-time:

```
1   //  Learning Sequence Control - Printing & Turning On LED
2   //  Use Hi-Res Timer to create specific time slots for various actions
3   //  Add first Println to clear the display screen and add 2nd Wait Time to see LED turn ON
4   START PROGRAM
5   {
6       Overall_Time  =  5.000sec
7       BT_Connect_Time  =  4.000sec
8       High-resolution Timer  =  Overall_Time
9       WAIT WHILE ( High-resolution Timer > BT_Connect_Time )
10      Print Screen with Line  =  1111
11      WAIT WHILE ( High-resolution Timer > 0.000sec )
12      Print Screen  =  Random Number
13      Print Screen with Line  =  Random Number
14      Aux LED (Blue)  =  TRUE (1)
15      High-resolution Timer  =  Overall_Time
16      WAIT WHILE ( High-resolution Timer > 0.000sec )
17  }
```

Fig. 3.2 Program “SequenceControl-2.tskx”.

Line 8 is executed and the Hi-Res Timer counter is loaded with Overall_Time (i.e. 5 seconds = 5000 ms). Next Line 9 is executed, i.e. the CM-50 is commanded to go to the Hi-Res Timer Counter and retrieve its current value which would be a few milliseconds less than 5000 (let’s say 4990), as some actual milliseconds had elapsed since the execution of line 8. Thus, the condition (Hi-Res Timer Counter > BT_Connect_Time) now becomes (4990 > 4000) and would evaluate to TRUE, therefore the CM-50 would be required to execute line 9 once more. So, this cycle of actions would continue for a while, but you can see that eventually this counter would be less than 4000, that is when about 1.0 second (± a few milliseconds) had elapsed since the execution of line 8. At this point, the condition (Hi-Res Timer Counter > BT_Connect_Time) evaluates to FALSE and the CM-50 can now quit Line 9 and executes Line 10 next, i.e. to print out “1111” along with a “new line”. In short, Lines 8-10 worked together to allow the printing of “1111” about one second after this program started on the CM-50. A more interesting way to look at these same actions is that we have programmed the CM-50 to allocate about 1.0 second for the proper BT connections to be made before it prints out “1111”. If for some reasons, BT connections could not be made within 1 second after the program started on the CM-50, we would not see “1111” printed out on the TASK Output Screen (see Video 3.4). As an exercise, the reader can try to allocate only 50 ms for BT connections (i.e. set BT_Connect_Time = 4950), and see if “1111” gets printed or not. The result would depend on the PC or mobile device that gets used.

The alert reader also may have noticed that no matter how many times this program was run, it always displayed the “same” two “random” numbers (lines 12 and 13).

Next, the reader can readily figure out that lines 14 to 16 work together to turn the blue LED on for 5 seconds. Although, we did not explicitly program the CM-50 to turn off the blue LED, when the CM-50 finished this program (i.e. executed line 17), it beeped and turned off the LED by itself.

The "SequenceControl-3.tskx" program applies the previous timing and sequence control schemes to the two motors on Ports 1 and 2 of the CM-50 (see Fig. 3.3).

```
1   // Learning Sequence Control - Turning on motors on Ports 1 and 2
2   // Use Hi-Res Timer to create specific time slots for various actions
3   // Add first Println to clear the display screen and add 2nd Wait Time to see turn ON motors
4   START PROGRAM
5   {
6       Overall_Time = 5.000sec
7       BT_Connect_Time = 4.000sec
8       High-resolution Timer = Overall_Time
9       WAIT WHILE ( High-resolution Timer > BT_Connect_Time )
10      Print Screen with Line = 1111
11      PORT[1]:Geared Motor = CCW:511 (49.95%)
12      WAIT WHILE ( High-resolution Timer > 0.000sec )
13      High-resolution Timer = Overall_Time
14      PORT[2]:Geared Motor = CCW:511 (49.95%)
15      WAIT WHILE ( High-resolution Timer > 0.000sec )
16  }
```

Fig. 3.3 Program "SequenceControl-3.tskx".

Lines 11 and 14 are LOAD statements which are used to control the direction and speed of each motor. Port 1 motor is programmed to turn CCW at a constant speed 511 for 9 seconds total, while Port 2 motor is programmed to turn CCW at a constant speed 511 for 5 seconds only. Video 3.5 shows how to modify "SequenceControl-2.tskx" into "SequenceControl-3.tskx" and documents how this program performs at run time. As before, when the CM-50 reaches the end of the program (line 16), it also stops all motors like it did previously for the blue LED. Please note that this ALL-OFF feature may not be implemented in other robotics kits from ROBOTIS.

Another way to look at the code in Fig. 3.3 is to treat a "WAIT WHILE" as a means for the Controller to be "inactive" during certain time periods (lines 9, 12 and 15), however during those same time periods the Robot was quite "active" as its actuators would still be executing the last commands given to them by the Controller (lines 11 and 14). **In other words, a "robot" does not have to be physically "inactive" whenever the "computer/controller" is set to be logically "inactive".** This "controlling/programming" scheme is used quite often in robotics and will be shown again in later examples in this book.

3.2 Learning Program Modularization with Functions in TASK programs

Let's start with the "Maneuvers-1.tskx" program (Fig. 3.4) which uses previous concepts already explained in Section 3.1. But this time, they are combined to make the robot go forward at Speed=512 for 0.5 s. and a COMPUTE type of statement is used to set separately the DIRECTION and SPEED for each of the Motor Ports 1 and 2 (lines 8 and 9 in Fig. 3.4).

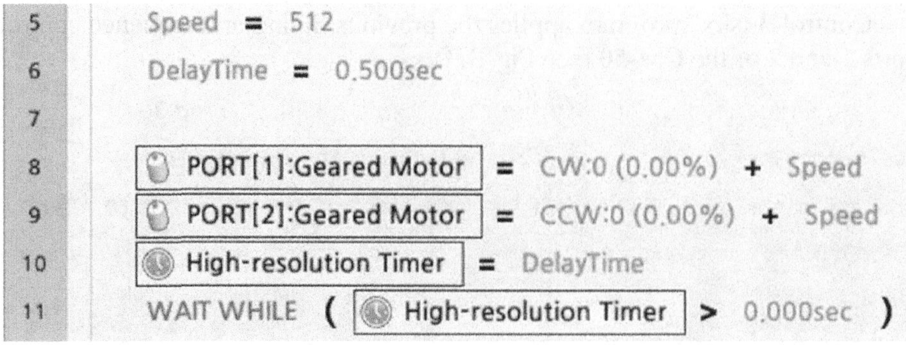

Fig. 3.4 Program "Maneuvers-1.tskx".

The COMPUTE statement is of the form "A = B *Op* C" which has the now familiar Assignment operator " = " separating the RHS expression (i.e. "B *Op* C") and LHS (i.e. "A"). The RHS now has two Operands "B" and "C" linked by a second operator *Op* which can be of 7 types:

" + " standing for the arithmetic addition operation (used in lines 8 and 9).

" - " standing for the arithmetic subtraction operation.

" * " standing for the arithmetic multiplication operation.

" / " standing for the arithmetic division operation.

" % " standing for the remainder operation (i.e. yields the "remainder" when doing "manual" integer division.

" & " standing for a bit-wise logical AND operation.

" | " standing for a bit-wise logical OR operation.

A "COMPUTE" statement is used here (line 8 of Fig. 3.4) instead of a "LOAD" statement (like in line 11 of Fig. 3.3) because SPEED is now a Variable Parameter, currently set to 512 but it can be changed to something else in other parts of the computer program (to be shown later in this Section 3.2). If the need was for a fixed numerical constant that would not change throughout the robot program, then a "LOAD" statement would be more efficient to use. Video 3.6 shows how to edit COMPUTE statements inside the program "Maneuvers-1.tskx", how this program perform at run-time and how to modify it into "Maneuvers-2.tskx" which uses a new type of computer programming structure called "FUNCTION" (see Fig. 3.5).

```
START PROGRAM
{
    Speed  =  512
    DelayTime  =  0.500sec

    PORT[1]:Geared Motor  =  CCW:0 (0.00%)  +  Speed
    PORT[2]:Geared Motor  =  CW:0 (0.00%)  +  Speed
    High-resolution Timer  =  DelayTime
    WAIT WHILE  ( High-resolution Timer  >  0.000sec )
}

FUNCTION  Forward
{
    PORT[1]:Geared Motor  =  CW:0 (0.00%)  +  Speed
    PORT[2]:Geared Motor  =  CCW:0 (0.00%)  +  Speed
    RETURN
}
```

Fig. 3.5 Program "Maneuvers-2.tskx".

The later part of Video 3.6 showed how to use the MAIN PROGRAM area as a "sandbox" area where the user can try out different coding ideas, while the FUNCTION areas were used as a repository of coding ideas that had worked well. This is a "shallow" understanding/application of the FUNCTION structure which serves as a useful but intermediary step towards a more complete understanding of the issues later.

Once the basic Functions "Forward", "Backward". "Right" and "Left" are defined, the program "Maneuvers-3.tskx" shows how to use them to create a sequence of robot maneuvers, perhaps needed to negotiate a maze, e.g. going forward, turn left then going forward again (see Fig. 3.6). These tasks were done by CALLing those predefined Functions "Forward" and "Left" as needed (lines 9, 13 and 17).

Part 1 of Video 3.7 explains the details of how the controller jumped from a CALL statement in the Main program (such as Line 9 in Fig. 3.6) into the Function definition area, executed the statements defined in this Function, then returned to the Main program to execute the next statement (e.g. Line 10).

```
4  START PROGRAM
5  {
6      Speed = 512
7      DelayTime = 0.500sec
8
9      CALL Forward
10     High-resolution Timer = DelayTime
11     WAIT WHILE ( High-resolution Timer > 0.000sec )
12
13     CALL Left
14     High-resolution Timer = DelayTime
15     WAIT WHILE ( High-resolution Timer > 0.000sec )
16
17     CALL Forward
18     High-resolution Timer = DelayTime
19     WAIT WHILE ( High-resolution Timer > 0.000sec )
20 }
```

Fig. 3.6 Program "Maneuvers-3.tskx".

The reader can see that by using Functions the logic flow in the Main program is easier to understand as it is presented in a more modular fashion, perhaps "logically coarser", while pushing the "gory" operational details into the Function definitions themselves. For example, this CM-50 robot has only two motors, but let's say that you move up to the DREAM system so your robot now has four motors to control instead of two, and you still need to perform the same "mission" (forward/left/forward). In this case, you would not have to modify the main program at all, but you will need to add new statements to take care of Motors 3 and 4 inside the appropriate Functions "Forward", "Backward", "Left" and "Right" (see Part 2 of Video 3.7).

Another useful application of Functions is to reduce the number of groups of repeating statements in the Main program. Fig. 3.6 shows three groups of the "same" statements involving the Hi-Res Timer (e.g. lines 10-11, 14-15 and 18-19). The program "Maneuvers-4.tskx" shows how to define the Function DELAY (Fig. 3.7) just one time and how to CALL it in the Main program as many times as needed (Fig. 3.8). "Maneuvers-4.tskx" also shows how to use the Variable Parameters Speed and DelayTime to create maneuvers with varying speeds and delay times, i.e. a more adaptive algorithm to deal with complicated mazes.

```
FUNCTION Delay
{
    High-resolution Timer = DelayTime
    WAIT WHILE ( High-resolution Timer > 0.000sec )
    RETURN
}

FUNCTION Stop
{
    PORT[1]:Geared Motor = CCW:0 (0.00%) + 0
    PORT[2]:Geared Motor = CCW:0 (0.00%) + 0
    RETURN
}
```

Fig. 3.7 Delay and Stop Functions in Program "Maneuvers-4.tskx".

```
    // Going Forward for 0.5 s at Speed 512
    Speed = 512
    CALL Forward
    DelayTime = 0.500sec
    CALL Delay

    // Turn right for 0.25 s at Speed 256
    Speed = 256
    CALL Right
    DelayTime = 0.250sec
    CALL Delay

    // Go Backward for 0.35 s at Speed 400
    Speed = 400
    CALL Backward
    DelayTime = 0.350sec
    CALL Delay

    // Stop robot before ending program
    CALL Stop
```

Fig. 3.8 Main Part of Program "Maneuvers-4.tskx".

Fig. 3.8 also shows a good coding practice of setting the value for a parameter needed for a given function, just before calling that function: for example, set up Speed (line 8) before calling Function Forward (line 9). Video 3.8 shows the runtime performance of the program "Maneuvers-4.tskx".

In this Section 3.2, the robot's maneuvers were "timed maneuvers", thus they may not work consistently for different/changing environments such as different ground surfaces that the robot may run on or the charge level of its batteries (low or high). In later sections, we'll use more "conditional" programming techniques so that the same robot code can work in different environments than the one that the code was originally created in.

3.3 "Spinning Top" Maneuvers using IR Center Sensor

We are now ready for our first implementation of the Sense-Think-Act paradigm previously discussed in Chapter 1. The sensing component will be the IR Center Sensor of the "Spinning Top" robot as shown in Fig. 3.9.

Fig. 3.9 "Spinning Top" robot and its IR Center Sensor.

The tasks assigned to this robot are as follows:

1) At start, the robot plays a musical melody to let the user know that it is "ready for action".

2) Next, the behaviors of the robot depend on the position of an object relative to the IR Center Sensor, thus the condition-action list to be considered is as follows:

If the IR Center Sensor reading is less than 50, the robot stays still.

If the IR Center Sensor reading is between 50 and 300, the robot spins right with its speed proportional to the IR Center Sensor reading and it plays a musical scale for 0.1 second.

If the IR Center Sensor reading is between 300 and 500, the robot spins left with its speed proportional to the IR Center Sensor reading and it plays a different second musical scale for 0.1 second.

If the IR Center Sensor reading is larger than 500, the robot rolls forward at Speed=512 to escape from the object and it plays a third musical scale for 0.5 second.

This condition-action list indicates that we are dealing with a situation whereas one and only one condition can happen at any one time during the runtime performance of this robot, thus we will use the IF-ELSE-IF structure for our solution to this problem. This challenge will be solved in two ways:

1) Using the built-in musical resources of the CM-50.

2) Using the text-display and musical functions available on a smartphone via the R+m.PLAY700 App.

3.3.1 Solution using CM-50's built-in speaker

The program "IR-based-Maneuvers.tsk" describes the solution approach used:

1) The Initial Event is solved by playing Melody 2 out of the CM-50 speaker (see Fig. 3.10). First, "Melody Time" is assigned to the "Buzzer Timer" (line 5), next the "Buzzer Index" is assigned with the value "Melody 2" (line 6). Then a WAIT WHILE statement is used to let the melody plays itself out (line 7). Lastly, Parameter Speed is initialized to zero (line 8).

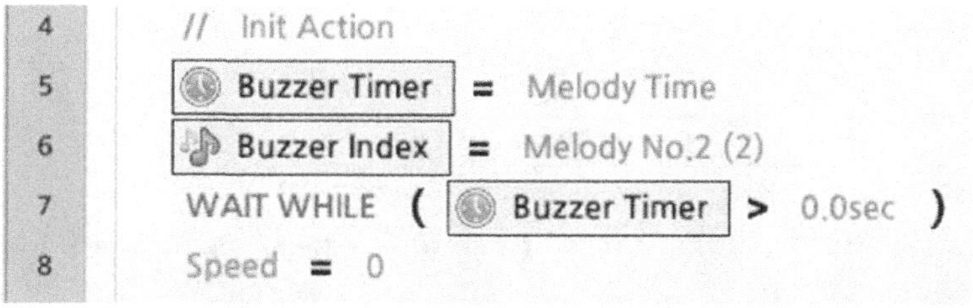

Fig. 3.10 Initial Event for program "IR-based-Maneuvers.tskx".

2) The challenge's algorithm (Fig. 3.11) uses an outer endless loop to represent the World-Robot Interactions Cycle (previously depicted in Fig. 1.10). This loop contains an IF-ELSE-IF structure representing the condition-action/input-output listing as detailed out previously.

3) During each loop, the controller would evaluate/execute this IF-ELSE-IF structure in the following manner:

It acquires the latest IR Center Sensor reading and saves it in Parameter IR_Sensor (line 13).

Next, it enters the IF-ELSE-IF structure and checks the first IF statement (line 14), i.e. the condition (IR_Sensor <= 50). If this condition is TRUE, the Function Stop is called (i.e. the robot stays still) and the controller will exit the whole IF-ELSE-IF structure at Line 41 and then will perform another loop from line 13. But if this condition is FALSE, i.e. (IR_Sensor > 50), the controller goes on to check the next ELSE-IF statement (line 18).

```
11 ENDLESS LOOP
12 {
13     IR_Sensor = IR Center
14     IF ( IR_Sensor <= 50 )
15     {
16         CALL Stop
17     }
18     ELSE IF ( IR_Sensor <= 300 )
19     {
20         Speed = IR_Sensor
21         CALL Right
22         Buzzer Timer = 0.1sec
23         Buzzer Index = Do# (4)
24         WAIT WHILE ( Buzzer Timer > 0.0sec )
25     }
26     ELSE IF ( IR_Sensor <= 500 )
27     {
28         Speed = IR_Sensor
29         CALL Left
30         Buzzer Timer = 0.1sec
31         Buzzer Index = Re# (6)
32         WAIT WHILE ( Buzzer Timer > 0.0sec )
33     }
34     ELSE IF ( IR_Sensor > 500 )
35     {
36         Speed = 512
37         CALL Forward
38         Buzzer Timer = 0.5sec
39         Buzzer Index = Fa# (9)
40         WAIT WHILE ( Buzzer Timer > 0.0sec )
41     }
42 }
```

Fig. 3.11 Main Algorithm used in program "IR-based-Maneuvers.tskx".

Please note that according to the considered condition-action listing, the controller is supposed to check if IR_Sensor is between 50 and 300 at line 18, but it uses a condition that involves only checking for (IR_Sensor <= 300). Is that sufficient to do what is required? And the answer is that it is OK to just check for the condition (IR_Sensor <= 300) here, because the condition (IR_Sensor > 50) was implicitly TRUE when line 18 was chosen (see previous paragraph). Thus if (IR_Sensor <= 300) turns out to be TRUE (meaning that IR_Sensor is between 50 and 300), the controller assigns the value of IR_Sensor to Speed and calls the Function Right to make the robot spin right at a Speed proportional to IR_Sensor's value. The controller also plays Do# for 0.1 second, then exits this IF-ELSE-IF structure and makes another loop starting at line 13. However, if the condition (IR_Sensor <= 300) turns out to be FALSE, i.e. (IR_Sensor > 300), then the controller goes on to execute the next ELSE-IF statement at line 26.

Similarly, executing line 26 means that the controller is checking to see if IR_Sensor is between 300 and 500. If this is TRUE, then the robot spins left with a Speed proportional to IR_Sensor's value, plays Re# for 0.1 second, exits this IF-ELSE-IF structure and continues with its looping task starting at line 13. If the condition in line 26 turns out to be FALSE (i.e. (IR_Sensor > 500), then line 34 is executed next.

Let's have a pop-quiz for line 34. **Was it necessary to check for (IR_Sensor > 500) on line 34? Or would a simple ELSE would do the job here?** The author would leave it to the readers to experiment on their own here. If (IR_Sensor > 500) turns out to be TRUE, then the controller would run forward at Speed=512 and for 0.5 second while playing Fa#. This maneuver is designed to make the robot run away from under the object. Once away from the object's influence, the IR_Center Sensor would then read zero, and the robot would stop and be ready to "play again" (i.e. lines 14-17).

Video 3.9 is a verbal explanation of the program "IR-based-Maneuvers.tskx" and showed the runtime performance of the "Spinning Top" robot.

3.3.2 Solution using Smartphone services: Text-Display, Text-to-Speech and Music-Play

In this section, the program "IR-based-Maneuvers.tskx" is adapted to make use of the Text-Display, Text-to-Speech and Instrument-Play services available on smartphones and tablets which are herein used as advanced remote devices. On TASK programs, these multimedia services are accessible via the SMART DEVICE and SMART CONSTANT categories and the resulting program is called "IR-based-Maneuvers-App.tskx". The user also needs to prepare a matching PLAY700 project on the Remote Device using the R+m.PLAY700 App (see Part 1 of Video 3.10), whereas 5 Text Items are set up (See Fig. 3.12).

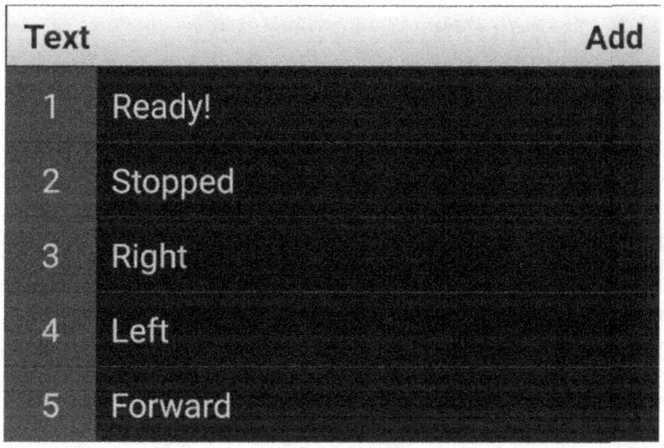

Fig. 3.12 Five Text Items used by program "IR-based-Maneuvers-App.tskx".

As far as TASK programming is concerned, the display screen of the remote device is divided into 25 zones (see Fig. 3.13). The top left zone is referred as Position (1,1), and the center zone is labeled as Position (3,3).

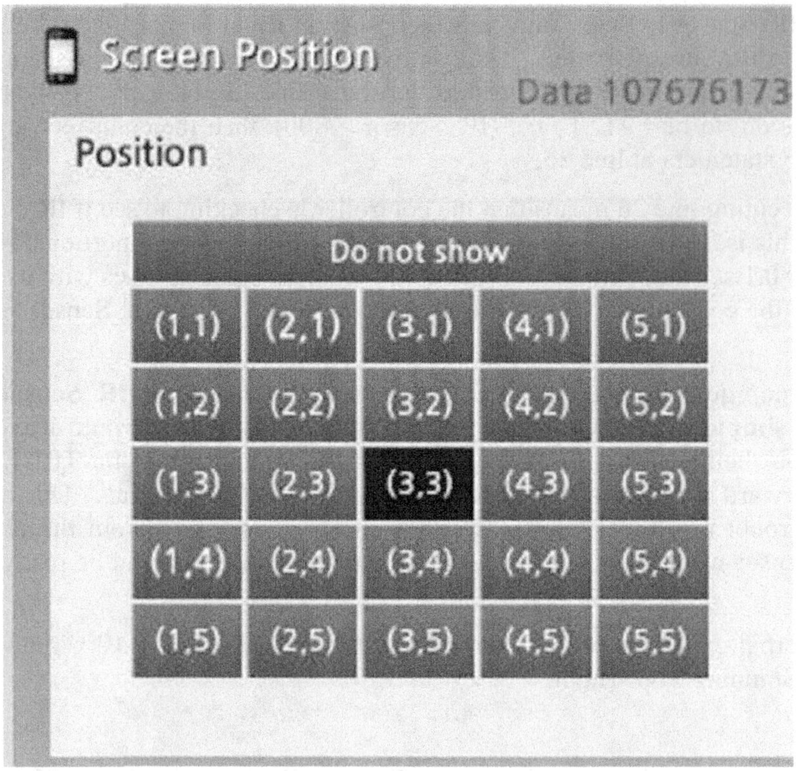

Fig. 3.13 The 25 zones of the Display Screen of Smart Devices.

The program "IR-based-Maneuvers-App.tskx" starts out in the same way as shown in Fig. 3.10 (i.e. it plays Melody 2 out of the CM-50's speaker). But the next actions differ because we are now dealing with the issue of waiting on the Bluetooth connections to be made securely between the Remote Device (i.e. R+m.PLAY700 App) and the Robot (i.e. TASK program). Fig. 3.14 shows the next group of statements:

Lines 10 and 11 set up the controller to listen for any Sound Clap before quitting the WAIT WHILE statement in line 11.

Thus, once the user creates some loud noise, the controller executes line 12 to set the Remote Device's Display into landscape mode.

Line 13 clears the Display, and line 14 displays Text Item 1 (i.e. "Ready!) at position (3,3) (i.e. at the Display's center), using a Text Size of 100 and White for Text Color.

Line 15 uses the Text-to-Speech engine to voice the Text Item 1 "Ready!". Line 16 uses a WAIT WHILE loop to wait for the speech engine to finish its job before continuing with the rest of the code. Please note that the Text-to-Speech engine cannot make the difference between "Ready" and "Ready!" at all (i.e. no intonation implemented).

```
9   // User needs to make sure that robot is connected to mobile device. User claps hands when ready.
10  Result of Sound Counter = 0
11  WAIT WHILE ( Result of Sound Counter == 0 )
12  SMART: Screen Rotation = Landscape Mode (2)
13  SMART: Text Display = 0
14  SMART: Text Display = [Position:(3,3)],[Item:1],[Size:100],[Color:White]
15  SMART: Text to Speech (TTS) = Text Item 1
16  WAIT WHILE ( SMART: Text to Speech (TTS) != 0 )
```

Fig. 3.14 Setting up Smart Device to display and voice out Text Item 1,

Fig. 3.15 shows the beginning of the Main Endless Loop in charge of monitoring the IR Center Sensor and deciding on the appropriate behavior(s) for the robot to perform, depending on the real-time value of that sensor:

Similarly, as before, the controller reads the IR Center Sensor and saves the value found there into Parameter "IR_Sensor" (line 20).

Then it enters the IF-ELSE-IF structure at line 21 to figure out if the object is present but still a way from the NIR Sensor itself (i.e. IR_Sensor <= 50). If this is TRUE, it calls Function Stop (line 23) and displays Text Item 2 (e.g. "Stopped") on the Display Screen's center in Red with a letter size of 100 (line 24). If this condition turns out to be FALSE (i.e. IR_Sensor > 50), then it would transfer to line 26 to check on the next ELSE-IF statement (see Fig. 3.16).

```
18  ENDLESS LOOP
19  {
20      IR_Sensor = IR Center
21      IF ( IR_Sensor <= 50 )
22      {
23          CALL Stop
24          SMART: Text Display = [Position:(3,3)],[Item:2],[Size:100],[Color:Red]
25      }
```

Fig. 3.15 Setting up Smart Device to display and voice out Text Item 2

Lines 26 to 33 in Fig. 3.16 describe the case when IR_Sensor is between 50 and 300, whereas the robot would spin right (line 29) and make the Smart Device display Text Item 3 (i.e."Right") in Green (line 30) and play the Do# note on the Marimba Instrument (line 31). A WAIT WHILE is used in line 32 to make sure that the Marimba finished its Do# note before getting out of this IF-ELSE-IF structure and performing another loop starting at line 20.

Lines 24 to 40 in Fig. 3.16 performed similar tasks for the case when IR_Sensor is between 300 and 500: i.e. spin left, display "Left" and play Re# on the Marimba.

Fig. 3.17 describes the last case when (IR_Sensor > 500). This is when the Remote Device displays Text Item 5, i.e. "Forward", and plays the Fa# note on a Trumpet (lines 48-49), while the robot rolls forward for 0.5 second (lines 44-46 and 50).

```
ELSE IF ( IR_Sensor <= 300 )
{
    Speed = IR_Sensor
    CALL Right
    SMART: Text Display = [Position:(3,3)],[Item:3],[Size:100],[Color:Green]
    SMART: Play a musical instrument = [Instrument:Marimba],[Octave:3],[Musical Scale:Do#(2)]
    WAIT WHILE ( SMART: Play a musical instrument > 0 )
}
ELSE IF ( IR_Sensor <= 500 )
{
    Speed = IR_Sensor
    CALL Left
    SMART: Text Display = [Position:(3,3)],[Item:4],[Size:100],[Color:Light gray]
    SMART: Play a musical instrument = [Instrument:Marimba],[Octave:3],[Musical Scale:Re#(4)]
    WAIT WHILE ( SMART: Play a musical instrument > 0 )
}
```

Fig. 3.16 Setting up Smart Device to respectively display Text Items 3 and 4 and play Do# and Re# on a Marimba while the robot spins left or right as appropriate.

```
42  ELSE IF ( IR_Sensor > 500 )
43  {
44      Speed = 512
45      CALL Forward
46      High-resolution Timer = 0.500sec
47      SMART: Text Display = [Position:(3,3)],[Item:5],[Size:100],[Color:Yellow]
48      SMART: Play a musical instrument = [Instrument:Trumpet],[Octave:3],[Musical Scale:Fa#(7)]
49      WAIT WHILE ( SMART: Play a musical instrument > 0 )
50      WAIT WHILE ( High-resolution Timer > 0 )
51  }
```

Fig. 3.17 Setting up Smart Device to display Text Item 5 and play Fa# on a Trumpet while the robot rolls forward for 0.5 second.

Part 2 of Video 3.10 shows the real-time performance of this program "IR-based-Maneuvers-App.tskx". **The reader may notice a small lack of responsiveness as compared to the previous solution "IR-based-Maneuvers.tskx", as this multimedia solution requires lots of Bluetooth communication traffic between the robot and the Smart Device.**

3.4 "Avoider" Maneuvers using Left and Right IR Sensors

In this section, we'll use the same robot platform as shown in Fig. 3.9, but it is renamed "Avoider" to emphasize the types of maneuvers described herein. The "Avoider" will be using its Left and Right IR Sensors to detect a single object and to accomplish tasks described in the following condition-action list:

If there is no obstacle in front of Avoider, it can go forward.

If there is an object on the Left of Avoider, it should back away and turn to the Right to avoid the obstacle.

If there is an object on the Right of Avoider, it should back away and turn to the Left to avoid the obstacle.

If there is an object in front of Avoider, it should go straight back to avoid the obstacle.

At first analysis, this seems to be a familiar programming situation of choosing one and only one alternative among several possible alternatives, thus an IF-ELSE-IF structure should do an adequate job when combined with an Outer Endless Loop. The TASK solution is named "Avoider-1.tskx". It starts with the playing of Melody 2 (same code as shown in Fig. 3.10). The main algorithm is described in Fig. 3.18:

The Endless Loop starts out by reading in the latest values from the IR Left and Right Sensors and saves them respectively into Variable Parameters IR_Left (line 14) and IR_Right (line 15).

Next is an IF-ELSE-IF structure with 4 mutually exclusive cases:

At line 16, if the values of Parameters "IR_Left" <u>and</u> "IR_Right" are both less than 10 (i.e. no obstacle is in front of the robot), the robot can go forward and the controller would then exit the IF-ELSE-IF structure and restart another loop at line 14. If this condition is FALSE, i.e. when either "IR_Left" <u>or</u> "IR_Right" have values larger than 10, or in other words when either IR sensor sees an object, the controller would need to check further by executing the next ELSE-IF statement at line 20.

```
12 ENDLESS LOOP
13 {
14     IR_Left  =  IR Left
15     IR_Right  =  IR Right
16     IF  ( IR_Left  <  10 && IR_Right  <  10 )
17     {
18         CALL Forward
19     }
20     ELSE IF  ( IR_Left >= 100 && IR_Right  <  10 )
21     {
22         CALL RightBack
23     }
24     ELSE IF  ( IR_Right >= 100 && IR_Left  <  10 )
25     {
26         CALL LeftBack
27     }
28     ELSE IF  ( IR_Left >= 100 && IR_Right >= 100 )
29     {
30         CALL Backward
31     }
32 }
```

Fig. 3.18 Main Algorithm used in program "Avoider-1.tskx".

The condition ((IR_Left >=100) AND (IR_Right < 10)), used for the ELSE-IF statement at line 20, represents a situation where there is an object on the left of the robot and none of its right side. If this condition evaluates to TRUE, the robot would turn right by backing away to avoid the obstacle on its left (line 22). The controller next exits this IF-ELSE-IF structure, and loops again at line 14. If this condition evaluates to FALSE, then line 24 is executed next.

Similarly, the condition ((IR_Right >=100) AND (IR_Left < 10)), used for the ELSE-IF statement at line 24, represents a situation where there is an object on the right of the robot and none of its left. If this condition evaluates to TRUE, the robot would back away and turn left to avoid the obstacle on its right (line 26). The controller next exits this IF-ELSE-IF structure, and loops again at line 14. If this condition evaluates to FALSE, then line 28 is executed next.

The ELSE-IF statement at line 28 uses this condition ((IR_Left >=100) AND (IR_Right >= 100)) to represent a situation when there is an object close enough to the robot's front such that both IR

Sensors report data values larger than 100. If this condition is TRUE, the robot would roll straight back away from the obstacle, otherwise the controller loops back to line 14.

Video 3.11 shows the run-time performance of this algorithm where the robot handled well the situations when the object was on either side of itself. But it did not handle well the situation when a "flat" object was right in front of it, as it kept on going back and forth at the same spot, i.e. it got "stuck". **This is a side-effect of the current algorithm that resulted in the controller "bouncing" back and forth between the conditions represented by line 16 and line 28.**

This problem can be easily fixed by adding a turning maneuver to the current backward maneuver (line 30). The turn direction does not matter, the author happens to use a left turn in the revised solution "Avoider-2.tskx" (see Fig. 3.19, lines 30-31). Video 3.12 shows much improved performance for frontal objects and corners.

```
28  ELSE IF ( IR_Left >= 100 && IR_Right >= 100 )
29  {
30     CALL Backward
31     CALL LeftBack
32  }
```

Fig. 3.19 Adding Left Turn to Backward Maneuver in program "Avoider-2.tskx".

The next program "Avoider-3.tskx" offers a different solution technique to these maneuvering tasks. **This new approach still relies on the same condition/action listing as described at the beginning of this section 3.4, but it distills different behavior patterns from this list:**

1) We must identify a "default" action which in this case is "Going Forward".

2) We will be using several simple IFs in parallel instead of a single big IF-ELSE-IF structure like before. **We also need to do a deeper analysis of the condition/action list to retain only the "essential" condition/action pairs.**

3) We also **need to apply a Delay Time to each action made** (this part was not needed when the IF-ELSE-IF structure was used). This Delay Time can be tuned for optimal performance.

Fig. 3.20 shows the relevant section of the algorithm used:

Line 10 showed that a Delay Time of 100 ms was used (the reader should try different values such as 0, 50, 150 etc... to see the effect on the overall performance of the robot).

An ENDLESS LOOP is still used (line 13) and starts by updating the parameters IR_Left and IR_Right with the latest data input from the IR Left and Right Sensors (lines 15 and 16).

The first IF statement (lines 17-21) represents the condition when there is an object on the robot's left (line 17), then the robot would back up and turn right (as before) to steer away from the obstacle (line 19). A short delay (line 20) is necessary for the robot to stay in this "action" to make this algorithm work properly.

The second IF statement (lines 22-26) represents the condition when there is an object on the robot's right (line 22), then the robot would back up and turn left (as before) to steer away from the obstacle (line 24). A short delay (line 25) is also necessary for the robot to stay in this "action" to make this algorithm work properly.

Please note that the "default" maneuver is always programmed as the "Last Action" ("Going Forward" in this case – line 27).

```
9   Speed = 256
10  DelayTime = 100
11
12  // Main Loop
13  ENDLESS LOOP
14  {
15      IR_Left = IR Left
16      IR_Right = IR Right
17      IF ( IR_Left >= 100 )
18      {
19          CALL RightBack
20          CALL Delay
21      }
22      IF ( IR_Right >= 100 )
23      {
24          CALL LeftBack
25          CALL Delay
26      }
27      CALL Forward
28  }
```

Fig. 3.20 Alternate Maneuvering Algorithm using "Parallel IFs" in program "Avoider-3.tskx".

It is important to perform a "thought experiment" here to "preview" the possible results of this Endless Loop at run time:

Every time that the Endless Loop is executed, line 27 will be executed, i.e. the robot will always go forward (a little bit), regardless of everything else that may happen.

Regarding the two parallel IFs, during some execution cycle(s), none of them could be activated (i.e. when there is no obstacle), or one of them could be triggered (i.e. when the obstacle is on either side of robot), or both could be triggered (i.e. a frontal obstacle). The alert reader may have noticed that seemingly the case for "a frontal obstacle" was not dealt with in this algorithm, so how would the robot behave at run time when encountering a frontal obstacle? Watching Video 3.13 will show the reader that the robot "wiggles" backward to deal with a frontal obstacle, i.e. the Functions RightBack and LeftBack were alternatively called. **This effect also explains why a DelayTime=0 would not work for the robot, as it would be switching so fast between the Functions "RightBack", "LeftBack" and "Forward" (within the Endless Loop) such that the robot looks "paralyzed" from the outside.** By the way, the same thing happens to us humans when we multitask too much as nothing then gets achieved!

At this point, some of the more advanced readers may have wondered whether the IF statements used in lines 17 and 22 of Fig. 3.20 could be changed to LOOP WHILE statements? And would the robot still work OK for this challenge?

The program "Avoider-4a.tskx" implements this idea (see Fig. 3.21) and Video 3.14 shows that this idea did not work! What went wrong? Was it wrong to use LOOP WHILE for this challenge? Watching Video 3.14 carefully would help the reader figure the "error" made.

```
13  ENDLESS LOOP
14  {
15      IR_Left  =  IR Left
16      IR_Right  =  IR Right
17      LOOP WHILE ( IR_Left >= 100 )
18      {
19          CALL RightBack
20          CALL Delay
21      }
22      LOOP WHILE ( IR_Right >= 100 )
23      {
24          CALL LeftBack
25          CALL Delay
26      }
27      CALL Forward
```

Fig. 3.21 Incorrect use of "LOOP WHILE" in program "Avoider-4a.tskx".

As it turns out, the error was not because LOOP WHILE was used per se, but it was more because LOOP WHILE was implemented incorrectly, as we forgot to "feed" them with the latest IR sensor data. We were using the Variable Parameters IR_Left and IR_Right to control the loops and we forgot to update their values every time we went through the LOOP WHILE (as we were supposed to!). It was a simple change to get to the program "Avoider-4b.tskx" (see Fig. 3.22) which does work properly. Please look at lines 17 and 22 in Fig. 3.22 where the real-time data from the sensors' ports are now accessed and used to control the LOOP WHILE, not <u>copies</u> of their values, i.e. parameters IR_left and IR_Right, as before. Please also note that lines 15 and 16 are commented out as they are no longer needed. See Video 3.15 for a runtime performance of this algorithm.

```
13  ENDLESS LOOP
14  {
15      // IR_Left  = IR Left
16      // IR_Right = IR Right
17      LOOP WHILE ( IR Left >= 100 )
18      {
19          CALL RightBack
20          CALL Delay
21      }
22      LOOP WHILE ( IR Right >= 100 )
23      {
24          CALL LeftBack
25          CALL Delay
26      }
27      CALL Forward
28  }
```

Fig. 3.22 Correct Use of "LOOP WHILE" in program "Avoider-4b.tskx".

There is an interesting experiment that the reader can make with the previous different versions of the Avoider codes: create a maze using books or boxes and let the Avoider robot roam freely through it. You will be surprised with its behaviors depending on which TASK code you used, especially when your maze has a cul-de-sac.

3.5 "Follower" Maneuvers using Left and Right IR Sensors

To round out the reader's experience with using the Left and Right IR Sensors, let's apply a **"reverse logic"** on them, essentially let's make the robot follow an object when it detects something in front of it.

Let's first learn how to track a moving object, when it is detected. The corresponding condition/action list can be as follows:

If there is no object detected, the robot should stay still (default behavior).

If there is an object detected by the Right IR Sensor, the robot should swing right.

If there is an object detected by the Left IR Sensor, the robot should swing left.

Should we also plan for the case when an object is detected and located right in front of the robot? Should it "stop" or should it show that it already "locks in" on the object in some other way? From the preceding Avoider coding work, we should be able to let the robot bounce between its two actions of swinging left and right to show that it got the object in its sight, right?

The program "Follower-1.tskx" was derived from the previous discussions using LOOP WHILEs. The algorithm used is shown in Fig. 3.23:

Essentially, whenever any IR Sensor (Left or Right) picks up on an object (lines 15 or line 19), the robot was turned, respectively, towards the same side where that IR Sensor was located (lines 17 or 21). Please note that a low detection threshold value of "5" is used so that the robot can detect an object as "far" as possible (about 2 inches away)!

```
13  ENDLESS LOOP
14  {
15      LOOP WHILE ( IR Left >= 5 )
16      {
17          CALL Left
18      }
19      LOOP WHILE ( IR Right >= 5 )
20      {
21          CALL Right
22      }
23      CALL Stop
24  }
```

Fig. 3.23 Object tracking algorithm in program "Follower-1.tskx".

The default maneuver is at line 23 which is interestingly a "non-maneuver" as it keeps the robot still (CALL Stop)!

Video 3.16 shows the runtime performance of this algorithm. **Please note the effect of the object's size on how the robot would tract the object.**

Here, the reader is challenged to come up with another algorithm using parallel IFs instead, however this new algorithm should yield the same object tracking result. Don't forget to use a small Delay Time to make this approach work!

In the next "Follower" application, a new condition/action pair is added to the previous list:

When the object gets a bit closer to the robot (i.e. threshold value of 10 applied to both IR Sensors), the robot should get even closer to the object by going Forward. At this point during runtime, if the user then moves the object away from the robot at a proper speed, ones would see the effect of the robot "following" the object (at a short distance away).

Surprisingly, there are two possible solutions: one using a combination of LOOP WHILE and an IF-ELSE-IF structure (Follower-2.tskx), and another one using only an IF-ELSE-IF structure (Follower-3.tskx). Who knows may be the readers can find other solutions too!

The program "Follower-2.tskx" (see Fig. 3.24) uses a default maneuver (CALL Stop) and an outer LOOP WHILE which encompass an inner IF-ELSE-IF structure that determines the one appropriate action among 3 possible and mutually exclusive alternatives "Left", "Right" or "Forward".

```
13  ENDLESS LOOP
14  {
15      LOOP WHILE ( IR Left >= 5 || IR Right >= 5 )
16      {
17          IF ( IR Left >= 5 && IR Right < 2 )
18          {
19              CALL Left
20          }
21          ELSE IF ( IR Right >= 5 && IR Left < 2 )
22          {
23              CALL Right
24          }
25          ELSE IF ( IR Left >= 10 && IR Right >= 10 )
26          {
27              CALL Forward
28          }
29      }
30      CALL Stop
31  }
```

Fig. 3.24 Object tracking algorithm in program "Follower-2.tskx".

Or in everyday language:

When no obstacle is detected, the LOOP WHILE (lines 15-29) would be skipped and line 30 would be executed, i.e. the robot would stay still as its default behavior.

When either IR Sensor detects an object (line 15), the LOOP WHILE is entered and the controller would check further with the IF-ELSE-IF structure to see first if the IR Left Sensor was the only one detecting the object (threshold "5" exceeded in line 17) then the robot would go Left (line 19). But if "line 17" turns out to be FALSE, the controller would check next to see if the IR Right Sensor was the only one detecting the object (threshold "5" exceeded in line 21), then the robot would go Right (line 23). But if "line 21" turns out to be FALSE also, then the controller would check to see if the object is close to both IR Sensors (line 25). If "line 25" turns out to be TRUE (i.e. both IR Sensors exceed a threshold of "10"), then the robot would go Forward (line 27). But if "line 25" turns out to be FALSE, the controller reaches the end of the IF-ELSE-IF structure without determining a concrete action, so the controller would just loop back to line 15 to check on the condition of the LOOP WHILE "((IR_Left >= 5) OR

(IR_Right >= 5))". If this condition is still TRUE, it would search through the IF-ELSE-IF structure again. If this condition becomes FALSE, the controller finally quits the LOOP WHILE, next triggers line 30 to stop the robot and then redo the outermost ENDLESS LOOP at line 14 all over again.

Please note, once again, that when using LOOP WHILE, we need to use LIVE IR Sensor Data inputs in all our conditional statements, so that most recent data from the sensors are used. Please see Video 3.17 for the runtime performance of "Follower-2.tskx".

The algorithm used in the next program "Follower-3.tskx" is described in Fig 3.25. It is more straight forward so it is easier to explain:

It uses an outermost ENDLESS LOOP (as always) (lines 13 to 33) and the first actions by the controller is to go out to the IR Sensors Ports and to read in the latest IR values and save them in parameters IR_Left and IR_Right (lines 15 and 16).

Next is the familiar IF-ELSE-IF structure which now represents 4 mutually exclusive condition/action pairs for "Left", "Right", "Forward" or "Stop". Please note that the IF-ELSE-IF structure in the previous "Follower-2.tskx" handles only 3 condition/action pairs.

```
13  ENDLESS LOOP
14  {
15      IR_Left  =  IR Left
16      IR_Right  =  IR Right
17      IF ( IR_Left >= 5 && IR_Right < 2 )
18      {
19          CALL Left
20      }
21      ELSE IF ( IR_Right >= 5 && IR_Left < 2 )
22      {
23          CALL Right
24      }
25      ELSE IF ( IR_Left >= 10 && IR_Right >= 10 )
26      {
27          CALL Forward
28      }
29      ELSE
30      {
31          CALL Stop
32      }
33  }
```

Fig. 3.25 Object tracking algorithm in program "Follower-3.tskx".

If line 17 is TRUE, i.e. only the IR Left Sensor sees the object, so the robot turns left.

Otherwise, line 21 is next executed, and if line 21 turns out to be TRUE, i.e. only the IR Right Sensor sees the object, then the robot turns right.

Otherwise, line 25 is executed next, and if line 25 turns out to be TRUE, i.e. both IR sensors see the object up close, then the robot goes forward to get even closer to the object.

Lastly, when no condition turns out to be TRUE (ELSE in line 29), the robot executes its default behavior, i.e. staying still.

Video 3.18 shows the run-time performance of this program "Follower-3.tskx".

There was no discernable difference in performance between these two solutions that the author could see. In theory, "Follower-3.tskx" should be "faster" because it accesses the sensor ports only twice per each iteration of the overall Endless Loop, i.e. it uses less resources from the controller as compared to "Follower-2.tskx" which uses from 4 to 8 accesses to the sensor ports per each iteration of its overall Endless Loop. However, "Follower-2.tskx" can be more "accurate" in its actions as it always uses "fresh" data inputs (i.e. "there is no free lunch").

Please note that ELSE is used in line 29 of "Follower-3.tskx", and its use is "critical" here. As an exercise for the reader, try to comment out line 29 (i.e. to make "Call Stop" the default behavior) and you will find that the robot no longer works properly. But if you then **add proper Delay Times** (~100 ms) for the actions "Left", "Right" and "Forward", you will find that the robot is working properly again! (see "Follower-3a.tskx"). So once again, the robot seems to work better when it is slowed down an appropriate amount!

Recapping Sections 3.4 and 3.5, the author's goal is to show that programming "autonomous" behaviors is fun and challenging at the same time! Starting from the (now familiar) condition/action analysis, **three techniques can be used to create TASK codes and after some initial actions, they all used an outermost ENDLESS LOOP for the main algorithm:**

1) In Technique 1, for each iteration of the Endless Loop, latest readings from the used sensors were first recorded, then an IF-ELSE-IF structure was used to represent all possible but mutually exclusive condition/action pairs that corresponded to robot behaviors required to fulfill a given challenge. Example programs using Technique 1 are "IR_based_Maneuvers.tskx", "IR_based_Maneuvers_App.tskx", "Avoider-1.tskx", "Avoider-2.tskx", and "Follower-3.tskx".

2) In Technique 2, the sensors reading task is still required first, but the IF-ELSE-IF structure is replaced by "parallel" IFs. This technique requires a deeper analysis of the original condition/action list to identify a "default" behavior, and other condition/action pairs that can work "on their own" or "in concert". This technique also requires a short time delay incorporated into each robot action. Compared to Technique 1, Technique 2 usually results in a smaller number of statements used in the TASK code, meaning that the controller would take less time to do each iteration of the Endless Loop, and thus the robot may become more responsive to changes in the "World" (review Fig. 1.10). Example programs using this technique are "Avoider-3.tskx" and "Follower-1-IFs.tskx".

3) Technique 3 is a variation of Technique 2 and it uses "parallel" LOOP WHILEs instead of "parallel" IFs. It combines "Data Acquisition" tasks in its Conditional Statements, thus it uses the least number of statements, as compared to the other two Techniques. Example programs using Technique 3 are "Avoider-4c.tskx" and "Follower-1.tskx". "Follower-2.tskx" is a hybrid solution combining Techniques 3 and 1.

3.6 Remote Control Programming Techniques

Back in Section 1.1, "Instruction" and "Status" packets were conceptually described, and "RemoCon" packets were first mentioned. In this Section 3.6, a more detailed description of "RemoCon" packets would be provided, along with programming examples of different schemes for Remote Control of the "Spinning Top" robot:

1) Remote Control via the Virtual RC-100 Controller (from inside the TASK tool).
2) Remote Control using Touch Areas on a mobile device display via the R+m.PLAY700 App.
3) Remote Control using Speech Recognition via the R+m.PLAY700 App.
4) Remote Control using Tilt Sensor on a mobile device via the R+m.PLAY700 App.

3.6.1 Structure of a "Remocon" packet

A full "Remocon" packet is based on a 16-bit user-designed message wrapped up in a 6-byte packet (http://support.robotis.com/en/product/auxdevice/communication/rc100_manual.htm). As this book is for younger beginning programmers, we'll restrict ourselves to using only the lower 10 bits of this 16-bit message. Advanced readers should consult Chapter 7 of Thai (2017) for a more complete treatment of this topic.

Both Physical and Virtual Remote Controllers RC-100 (respectively, Figs 1.5 and 1.6) were designed to behave programmatically in the same manner:

Each of the 10 buttons U-D-L-R-1-2-3-4-5-6 is represented by 1 bit in a 10-bit number/message, with the "U" button using the bit-0 position and the "6" button using the bit-9 position (see Fig. 3.26).

RC-100 10-bit Message

Bit Position	*9*	*8*	*7*	*6*	*5*	*4*	*3*	*2*	*1*	*0*
Button	6	5	4	3	2	1	R	L	D	U
Example Data	0	0	0	1	0	0	1	0	0	1

Fig. 3.26 Structure of an RC-100 10-bit message.

Whenever one or several buttons are pushed by the user, this 10-bit message is sent from the RC-100 to the robot. The "Example Data" row in Fig. 3.26 shows that Buttons "U", "R" and "3" are pushed as these bits are set to "1". All the "un-pushed" buttons of course have their bits set to 0.

Lastly, whenever the user releases a button, the RC-100 would send the number "0" (i.e. all bits set to 0) ONCE. This feature is unique with the ROBOTIS RC-100, other manufacturers may not implement the same scheme.

The main issues in using the RC-100 are to know how many buttons are being pushed and how to sort out which buttons are being pushed as information about all 10 buttons is sent inside each RC message.

First, the issue of "how many" is controlled by the programmer, as readers may well know that it is impossible to expect that actual users of one's programs will only push the buttons that the programmer would want them to! The program "RC-1Button.tskx" (Fig. 3.27) shows how to restrict one's program to deal only with 1 "expected" button ("U" or "D") and also to ignore any other situation.

After the usual Initial Action section whereas Melody 2 is played and Variables Data and DelayTime are initiated respectively to 0 and 250 (i.e. 250 ms), the main algorithm is shown in Fig. 3.27:

```
12  ENDLESS LOOP
13  {
14      WAIT WHILE ( Remocon Data Received == FALSE (0) )
15      Data = Remocon RXD
16      Print Screen with Line = Data
17      IF ( Data == U )
18      {
19          Print Screen = Data
20          Print Screen with Line = 11
21      }
22      ELSE IF ( Data == 0000 0000 0000 0000 0000 0000 0000 0010 )
23      {
24          Print Screen = Data
25          Print Screen with Line = 22
26      }
27      CALL Delay
28  }
```

Fig. 3.27 Main algorithm in program "RC-1Button.tskx".

Statements 14 and 15 need to be used together and they represent the first actions to be done in a basic RC Loop. The CM-50 Controller's Firmware has a built-in procedure that keeps on listening on the RC communication channel and if a new message has been received, it sets the "Remocon Data Received" flag to TRUE, otherwise this flag is set to FALSE. Thus, at each iteration of the Endless Loop (lines 13-28), the CM-50 would first check the WAIT WHILE's condition set in line 14. This condition evaluates to (FALSE == FALSE) if no button has been pushed, thus this WAIT WHILE's condition evaluates to TRUE, and the CM-50 would stay at line 14. Eventually, the user pushes one or many buttons on the RC-100. This would make the flag "Remocon Data Received" become TRUE and the WAIT WHILE's condition now becomes (TRUE == FALSE) which evaluates to FALSE. When this happens, line 14 is exited and line 15 is executed next. It goes to the memory buffer "Remocon RXD" and save the value found there into the Parameter "Data". Line 16 essentially prints out a decimal number that represents the status of all 10 buttons.

As this program only cares if only ONE button is pushed and furthermore this single button has to be either "U" or "D" (but not both), the IF-ELSE-IF structure (lines 17-26) implements these requirements. Line 17 checks to see if "Data" is equal to Button "U", if this turns out to be TRUE, the CM-50 prints "Data" and then the number "11" as a marker (lines 19-20). If line 17 turns out to be FALSE, then line 22 is executed next. Line 22 checks to see if "Data" is equal to Button "D", here Button "D" is shown as its binary-number equivalent (showing that Bit 1 is set). If line 22 turns out to be TRUE, the CM-50 prints "Data" and then the number "22" as a marker (lines 24-25). If line 22 turns out to be FALSE, then line 27 is executed next (i.e. a 250 ms delay so that the user can reasonably read the Output Window inside the TASK tool). Then another iteration of this Endless Loop begins again at line 14.

Video 3.19 shows a typical run of this program whereas all messages sent by the RC-100 were shown, but only those single-button ("U" or "D") messages were processed further, i.e. printed out with either marker "11" or "22" as appropriate.

The next program "RC-2Buttons.tskx" shows how to parse out two specific buttons "Button1" and "Button2" from the RC-100's 10-bit message. "Button1" needs to be a single-button input from the group "U-D-L-R", while "Button2" needs to be a single-button input from the group "1-2-3-4". The main algorithm is shown in Fig. 3.28 and it has two new features - a binary filtering procedure using the bit-level "&" operator (lines 20 and 22) and a combo-condition using the logical OR operator (i.e. "||") implemented in each of two parallel IFs structures:

The Endless Loop starts with the usual check of the "Remocon Data Received" flag (line 16), next saves the most recent RC-100 input message into Parameter Data (line 17) and prints it (line 18).

Button1 is extracted out of Parameter Data by ANDing it with the Mask U+D+L+R (line 20 in Fig. 3.28 and more details in Fig. 3.29). In this Figure, "Example Data" shows that Buttons "L" and "3" were pushed, but after the "&" operation with the Mask "U+D+L+R", only the information from Button "L" remains. More details are explained in Video 3.20.

Similarly, Button2 is extracted out of Parameter Data by ANDing it with the Mask 1+2+3+4 (line 22 in Fig. 3.28 and more details in Fig. 3.30). In this Figure, "Example Data" shows that Buttons "L" and "3" were pushed, but after the "&" operation with the Mask "1+2+3+4", only the information from Button "3" remains.

Next, two parallel IFs are used to check on Button1 and Button2 separately for their actual "values", i.e. membership within their appropriate groups/masks (lines 23 and 28). If each or both IFs turn out to be true, then the values for Button1 or/and Button2 would be printed along their appropriate markers "11" or "22".

```
ENDLESS LOOP
{
    WAIT WHILE ( Remocon Data Received == FALSE (0) )
    Data = Remocon RXD
  Print Screen with Line = Data
    // Filtering out U-D-L-R buttons for Button1
    Button1 = Data & U+D+L+R
    // Filtering out 1-2-3-4 buttons for Button2
    Button2 = Data & 1+2+3+4
    IF ( Button1 == U || Button1 == D || Button1 == L || Button1 == R )
    {
        Print Screen = Button1
        Print Screen with Line = 11
    }
    IF ( Button2 == 1 || Button2 == 2 || Button2 == 3 || Button2 == 4 )
    {
        Print Screen = Button2
        Print Screen with Line = 22
    }
    CALL Delay
}
```

Fig. 3.28 Main algorithm in program "RC-2Buttons.tskx".

AND Operation with Mask U-D-L-R

Bit Position	*9*	*8*	*7*	*6*	*5*	*4*	*3*	*2*	*1*	*0*
Button	**6**	**5**	**4**	**3**	**2**	**1**	**R**	**L**	**D**	**U**
Example Data	0	0	0	1	0	0	0	1	0	0
						&				
Mask U+D+L+R	**0**	**0**	**0**	**0**	**0**	**0**	**1**	**1**	**1**	**1**
Result from & Op.	***0***	***0***	***0***	***0***	***0***	***0***	***0***	***1***	***0***	***0***

A	B	A & B
1	0	0
1	1	1
0	0	0
0	1	0

Fig. 3.29 Result from ANDing operation between "Example Data" and Mask (U+D+L+R).

AND Operation with Mask 1-2-3-4

Bit Position	*9*	*8*	*7*	*6*	*5*	*4*	*3*	*2*	*1*	*0*
Button	**6**	**5**	**4**	**3**	**2**	**1**	**R**	**L**	**D**	**U**
Example Data	0	0	0	1	0	0	0	1	0	0
						&				
Mask 1+2+3+4	**0**	**0**	**1**	**1**	**1**	**1**	**0**	**0**	**0**	**0**
Result from & Op.	***0***	***0***	***0***	***1***	***0***	***0***	***0***	***0***	***0***	***0***

A	B	A & B
1	0	0
1	1	1
0	0	0
0	1	0

Fig. 3.30 Result from ANDing operation between "Example Data" and Mask (1+2+3+4).

Video 3.20 shows the performance of this program "RC-2Buttons.tskx" at run-time.

3.6.2 Remote Control of "Spinning Top" robot using Virtual RC-100

We are now ready to expand the concepts learned from Section 3.6.1 to work on a Remote- Control project for a CarBot which allows the user to use multiple Direction Buttons among the group U-D-L-R and to additionally set Low-Speed or High-Speed parameters on the fly using Buttons 1 and 3.

The first part of the main algorithm is shown in Fig. 3.31:

```
20  ENDLESS LOOP
21  {
22      WAIT WHILE ( Remocon Data Received == FALSE (0) )
23      Data = Remocon RXD
24      Print Screen with Line = Data
25      IF ( Data == -- )
26      {
27          CALL Stop
28      }
```

Fig. 3.31 First part of algorithm used in "RC-MultiDirectionsSpeeds.tskx".

After the usual "Remocon Data Received" flag checking and the reading/printing of the incoming message (lines 22-24), if the incoming Data turns out to be "0" (line 25), i.e. "No Button" is pushed, "CALL Stop" is used to stop all motions for the robot (line 27).

If line 25 turns out to be FALSE, i.e. some Buttons are pushed, then the ELSE branch is taken (line 29) and each possible Direction and Speed Buttons are parsed out using the "&" Operator with the appropriate masks (see Fig. 3.32, lines 32-37).

```
29      ELSE
30      {
31          // Filtering out specific U-D-L-R-1-3 buttons
32          ButtonUp = Data & U
33          ButtonDown = Data & D
34          ButtonLeft = Data & L
35          ButtonRight = Data & R
36          Button1 = Data & 1
37          Button3 = Data & 3
```

Fig. 3.32 Parsing out various Direction and Speed Buttons in "RC-MultiDirectionsSpeeds.tskx.

Next, 6 parallel IFs are used to trigger the appropriate maneuvers and speed settings depending on which Buttons are determined to be pushed at run-time (see Fig. 3.33).

Video 3.21 shows the runtime performance of the program "RC-MultiDirectionsSpeeds.tskx" where ones can observe that Speed settings can be made in parallel with the directional maneuvers of the robot. While the maneuvers themselves can be combined with each other in a functional way (e.g. "U" and "R" together) or in a completely ineffective way (e.g. "U" and "D" together)!

```
IF ( ButtonUp == U )
{
    CALL Forward
}
IF ( ButtonDown == D )
{
    CALL Backward
}
IF ( ButtonLeft == L )
{
    CALL Left
}
IF ( ButtonRight == R )
{
    CALL Right
}
IF ( Button1 == 1 )
{
    Speed = 200
}
IF ( Button3 == 3 )
{
    Speed = 300
}
```

Fig. 3.33 Parallel IFs used for setting Direction(s) and Speed in "RC-MultiDirectionsSpeeds.tskx".

3.6.3 Remote Control of "Spinning Top" robot using Touch Areas on Mobile Devices

This Section 3.6.3 shows how to write a TASK program that can be interfaced with a mobile device so that its "Touch Area" service can be utilized to perform Remote Control of a CarBot.

Let's first look at a basic application for using the Touch Area features offered by the R+m.PLAY700 mobile App. This mobile App divides the display area in a 5x5 grid with a total of 25 touch zones (see Fig. 3.34). This App can monitor **up to 2** Touch Areas simultaneously.

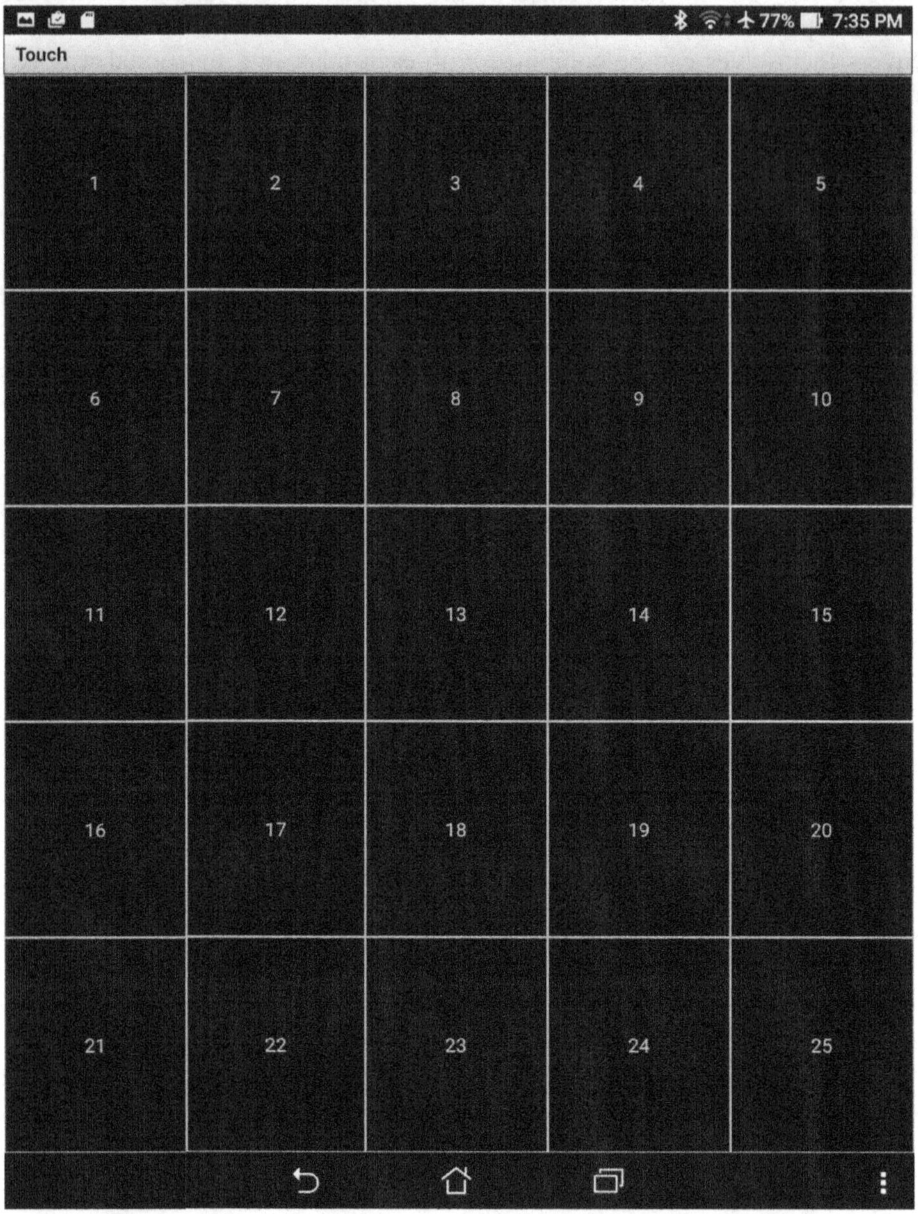

Fig. 3.34 The 25 Touch Areas on a mobile device's display using App R+m.PLAY700.

The program "MonitorTouchAreas.tskx" shows how to detect where the user had touched the mobile display and how to draw either a circle or a square in those detected zones.

First, we need to create a "Remote Control" project on the mobile device via the App R+m.PLAY700 so that we can assign 3 Shape Items (see Fig. 3.35 and part 1 of Video 3.22).

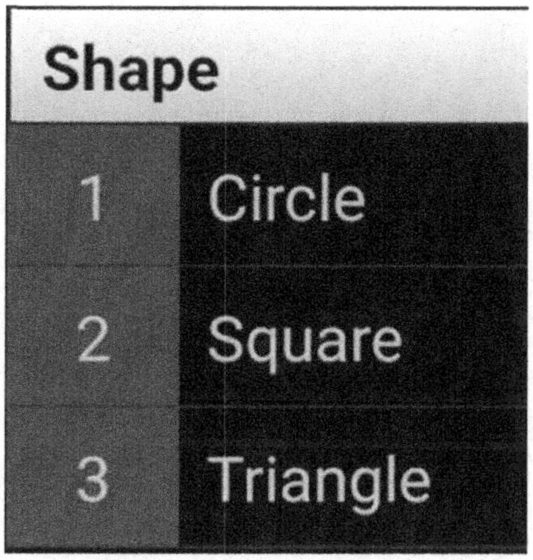

Fig. 3.35 Assigning 3 Shape Items using App R+m.PLAY700.

Fig. 3.36 shows the initial actions of the program "MonitorTouchAreas.tskx":

1) Set the mobile display to the Portrait Mode (line 6).

2) Clear the display of all previous shapes – for just in case (line 7).

3) Wait for the user to whistle into the microphone of the mobile device (line 9) before going into the main Endless Loop (line 11 of Fig. 3.37).

```
3  START PROGRAM
4  {
5      // Getting remote device ready using SMART commands
6      SMART: Screen Rotation = Portrait Mode (1)
7      SMART: Shapes Display = 0
8      // User needs to make sure that robot is connected to mobile device. User whistles loudy when ready.
9      WAIT WHILE ( SMART: Noise (dB) < 50 )
```

Fig. 3.36 Initial actions by program "MonitorTouchAreas.tskx".

```
11 ENDLESS LOOP
12 {
13     Touch1 = SMART: Touch Area 1
14     Touch2 = SMART: Touch Area 2
15     // Clearing screen display
16   SMART: Shapes Display = 0
17     IF ( Touch1 != 0 )
18     {
19       SMART: Shapes Display = [Position:Touch1],[Item:1],[Size:60],[Color:White]
20       High-resolution Timer = 0.250sec
21       WAIT WHILE ( High-resolution Timer > 0 )
22     }
23     IF ( Touch2 != 0 )
24     {
25       SMART: Shapes Display = [Position:Touch2],[Item:2],[Size:60],[Color:Green]
26       High-resolution Timer = 0.250sec
27       WAIT WHILE ( High-resolution Timer > 0 )
28     }
29 }
```

Fig. 3.37 Main Algorithm in program "MonitorTouchAreas.tskx".

Fig. 3.37 describes the main algorithm controlled by a familiar Endless Loop, with the following actions for each iteration of this loop:

1) in Lines 13 and 14, SMART commands are used to access the Touch Area 1 and 2 where the user may have touched the mobile display. If no touching is done by the user, a zero value is saved into Parameters "Touch1" and "Touch2", otherwise a number between 1 and 25 will be returned from those SMART commands. The "first" touch by the user will be recorded into "Touch1" and the "second" touch will be recorded into "Touch2".

2) The display screen is cleared of previous graphics (line 16).

3) Two parallel IFs are used, one for each Touch, as 1 or 2 touches can be performed by the user at any time. If "Touch1" is non-zero (line 17), then a Circle (i.e. Shape Item 1) will be displayed at the location defined by the value contained in "Touch1" in White color (line 19). Please see Fig. 3.38 for details of the RHS of the LOAD statement at line 19 of Fig. 3.37.

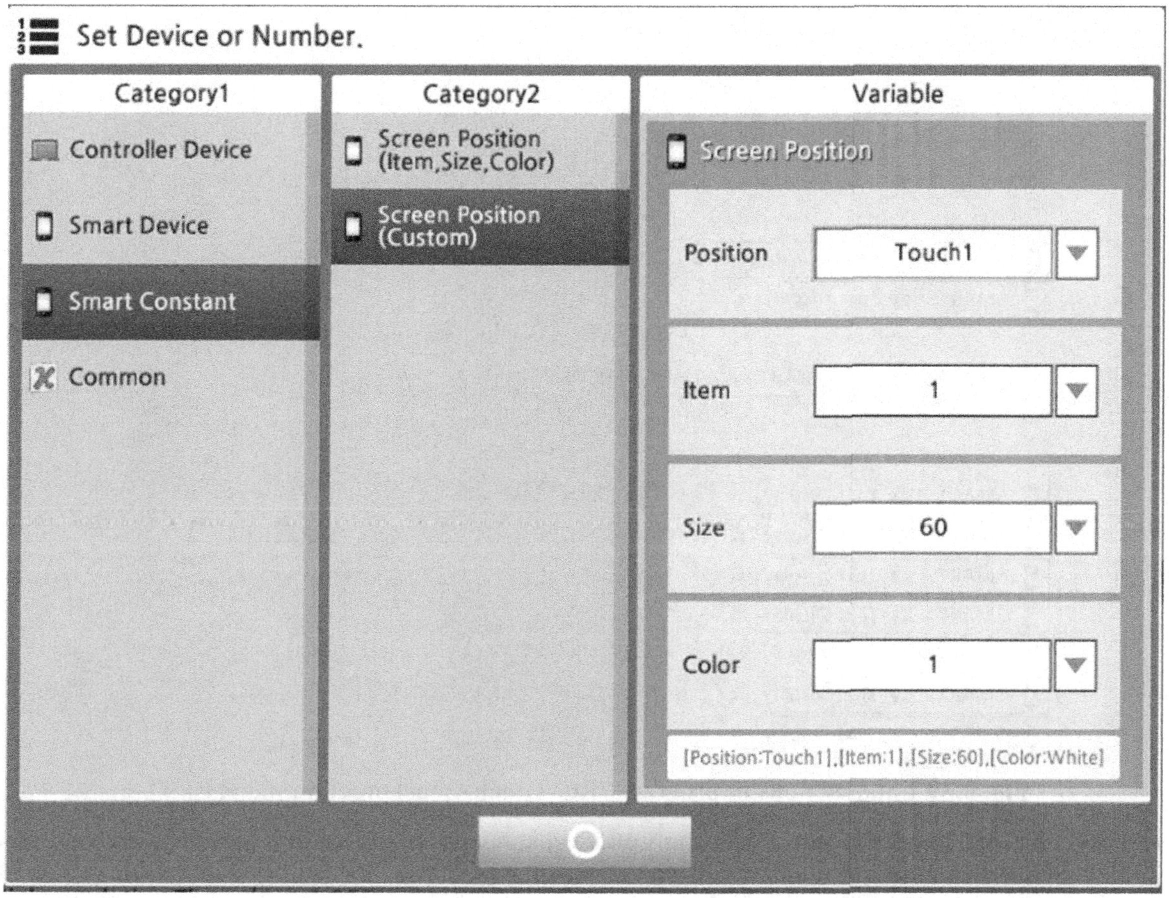

Fig. 3.38 Details of RHS of Load Statement at line 19 of Fig. 3.37.

4) Next is a Time Delay lasting 250 ms (lines 20-21) so that users can catch sight of the displayed shape.

5) Lines 23-28 shows a similar procedure used for Location "Touch2", but now a Green Square shape (i.e. Shape Item 2) is used.

Part 2 of Video 3.22 shows the runtime performance of the program "MonitorTouchAreas.tskx".

As we can only use up 2 Touch Areas simultaneously for this RC project, we cannot achieve the multi-button features allowed by the RC-100 as shown previously in Section 3.6.1. However, we still want to be able to set the CarBot's maneuver directions independently from its Speed setting. The solution approach is to use one "Touch" for the main maneuver direction and the other "Touch" for a secondary maneuver direction or for setting the Speed value.

The program "RC-TouchAreasDirectionSpeed.tskx" describes how this solution approach is implemented:

1) The initial actions are quite extensive in this solution (see Fig. 3.39). Three parameters are used "Speed" (= 256), "TimeDelay1" (= 100 ms) and "TimeDelay2" (= 25 ms) as shown in lines 5-7. Next the mobile device is set to Portrait mode, and its display cleared, then it waits on the user to whistle into it (line 12) to get the CM-50 to finish up its main actions.

```
3   START PROGRAM
4   {
5       Speed = 256
6       TimeDelay1 = 100
7       TimeDelay2 = 25
8       // Getting remote device ready using SMART commands
9       SMART: Screen Rotation = Portrait Mode (1)
10      SMART: Text Display = 0
11      // User needs to make sure that robot is connected to mobile device. User whistles loudy when ready.
12      WAIT WHILE ( SMART: Noise (dB) < 50 )
13
14      // Display Direction Control Text FORWARD, BACKWARD, LEFT, RIGHT
15      SMART: Text Display = [Position:(3,2)],[Item:1],[Size:100],[Color:White]
16      SMART: Text Display = [Position:(3,4)],[Item:2],[Size:100],[Color:White]
17      SMART: Text Display = [Position:(2,3)],[Item:3],[Size:100],[Color:White]
18      SMART: Text Display = [Position:(4,3)],[Item:4],[Size:100],[Color:White]
19      // Display Speed Control Text SLOW FAST
20      SMART: Text Display = [Position:(1,1)],[Item:5],[Size:100],[Color:Red]
21      SMART: Text Display = [Position:(5,1)],[Item:6],[Size:100],[Color:Red]
```

Fig. 3.39 Initial actions in program "RC-TouchAreasDirectionSpeed.tskx".

2) Next, six Text Items are put at the appropriate display locations so that they can be used as icons standing in for maneuver directions and speed settings (see Fig. 3.40 and lines 15-21 in Fig. 3.39). Essentially, these six Text Items (and their underlying Touch Zones) serve as the user interface for the program "RC-TouchAreasDirectionSpeed.tskx", similar in function to the Virtual RC-100 user interface.

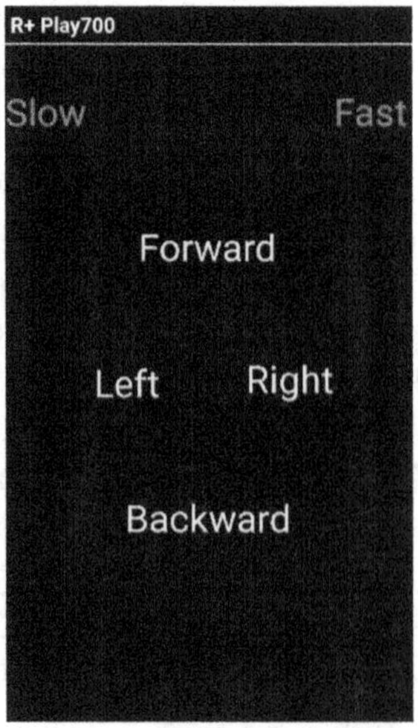

Fig. 3.40 Screen capture showing user interface in "RC-TouchAreasDirectionSpeed.tskx".

```
24  ENDLESS LOOP
25  {
26      Touch1  =  SMART: Touch Area 1
27      Touch2  =  SMART: Touch Area 2
28      IF ( Touch1 != FALSE (0) || Touch2 != FALSE (0) )
29      {
30          IF ( Touch1 == 8 || Touch2 == 8 )
31          {
32              CALL Forward
33              CALL Delay1
34          }
35          IF ( Touch1 == 18 || Touch2 == 18 )
36          {
37              CALL Backward
38              CALL Delay1
39          }
40          IF ( Touch1 == 12 || Touch2 == 12 )
41          {
42              CALL LeftTurn
43              CALL Delay2
44          }
45          IF ( Touch1 == 14 || Touch2 == 14 )
46          {
47              CALL RightTurn
48              CALL Delay2
49          }
```

Fig. 3.41 Part 1 of Control Algorithm used in "RC-TouchAreasDirectionSpeed.tskx".

3) The control solution is contained in a familiar Endless Loop (lines 24-63). For each iteration, the first part of the algorithm is shown in Fig. 3.41.

Parameters "Touch1" and "Touch2" record the current Touch Area 1 and Touch Area 2 as reported by the R+m.PLAY700 App (lines 26-27). "Touch1" and "Touch2" would read zero if no touch had been detected.

Line 28 shows that if either (or both) "Touch1" or "Touch2" is (are) non-zero (possible values are from 1 to 25), this means that the user has touched the mobile display somewhere. The next four parallel IF structures (lines 30-49) are used to associate appropriate Touch Areas to their corresponding Directional Maneuvers (Forward, Backward, Left and Right). The parallel IFs are used to allow the user to combine

2 different moves together if wanted (for example, Forward and Left). Please note that a Delay1 function (i.e. 100 ms delay) is used for Forward and Backward moves, while a Delay2 function (i.e. 25 ms delay) is used instead for Left and Right moves. Please also note the OR ("||") operator used in the Conditional Expression used in these 4 IF statements, as no one can really predict which Touch Area is the user is going to touch first (or secondly).

4) The second part of the algorithm is shown in Fig. 3.42, where an IF-ELSE-IF structure is used to take care of setting the appropriate speed levels "Slow" or "Fast" (i.e. 200 or 300) depending on whether Zone 1 or Zone 5 was touched. Please note that an IF-ELSE-IF structure is used here because "Slow" or "Fast" are two mutually exclusive alternatives.

Finally, line 59 (ELSE) is executed if line 28 evaluates to FALSE (i.e. no touch detected on the mobile display), then the Carbot is stopped (line 61).

```
50          IF ( Touch1 == 1 || Touch2 == 1 )
51          {
52              Speed = 200
53          }
54          ELSE IF ( Touch1 == 5 || Touch2 == 5 )
55          {
56              Speed = 300
57          }
58      }
59      ELSE
60      {
61          CALL Stop
62      }
63  }
```

Fig. 3.42 Part 2 of Control Algorithm used in "RC-TouchAreasDirectionSpeed.tskx".

Video 3.23 shows the run-time performance of this program "RC-TouchAreasDirectionSpeed.tskx".

3.6.4 Voice Control of "Spinning Top" robot using R+m.PLAY700 App

This project explores the feasibility and performance of a Voice Control scheme to remotely control a CarBot to perform basic maneuvers. Back in Section 3.3.2, Text Items were useful to display texts and could serve as "inputs" to the Text-to-Speech engine of the R+m.PLAY700 App. The same Text Items can also be used as "outputs" from this App's Speech Recognition engine which was based on the Google Speech API (https://cloud.google.com/speech/), meaning that your mobile device needs to be connected to the web in order to use this feature and consequently this process needs "seconds" to perform, not "milliseconds" like with the RC-100.

Another important observation also needs to be made regarding the difference in Remote Control "technologies" that are used so far. When using the RC-100 module, potentially our RC scheme could rely upon 10 simultaneous Input Sources as represented by the 10 Buttons "UDLR123456". When using the Touch Areas of a mobile display, we can only rely on two simultaneous Input Sources, Touch Areas 1 and 2 even though the potential number of touch zones is 25. In this Section 3.6.3, we plan to use Speech Recognition technology which potentially can recognize human verbal inputs, but it must be noted that we will be restricted to ONE verbal input stream at any one time, as humans do not know how to multiplex our speech pattern yet!

Fig. 3.43 shows the 10 Text Items used for this PLAY700 project (please review Video 2.2 regarding the use of the Speech/Voice Recognition tool as needed).

Fig. 3.43 Ten Text Items used with program "RC-UsingSpeech.tskx".

The solution to this project is provided by the program "RC-UsingSpeech.tskx". Fig. 3.44 describes the initial Actions to be performed:

Variable Parameters "Speed", "TurnSpeed", "VoiceCommand" and Flag "InvalidCommand" get initialized by lines 5-8.

Then if the user whistles into the mobile device (line 10), line 11 is executed next to display on the mobile device a Menu of valid commands that can be voiced out by the user (see Fig. 3.45).

```
START PROGRAM
{
   Speed  =  200
   TurnSpeed  =  160
   VoiceCommand  =  Not Used (0)
   InvalidCommand  =  FALSE (0)
   // User needs to make sure that robot is connected to mobile device. User whistles loudy when ready.
   WAIT WHILE  ( SMART: Noise (dB)  <  50 )
   CALL  Menu
```

Fig. 3.44 Initial Actions in program "RC-UsingSpeech.tskx".

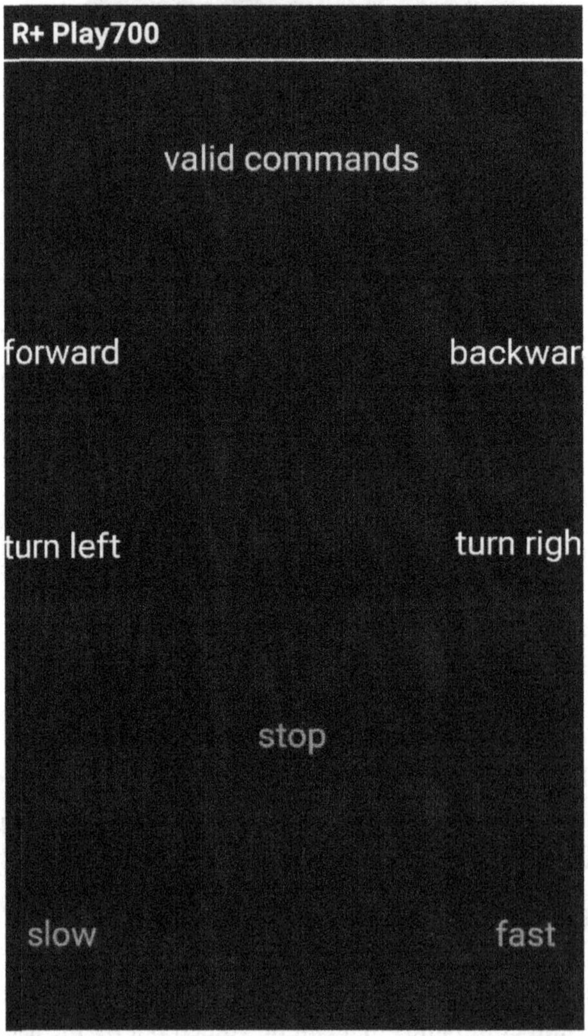

Fig. 3.45 Menu of Valid Verbal Commands in program "RC-UsingSpeech.tskx".

```
13  ENDLESS LOOP
14  {
15  Whistle :
16     WAIT WHILE ( SMART: Noise (dB) < 50 )
17     SMART: Text Display = 0
18     SMART: Text to Speech (TTS) = Text Item 1
19     SMART: Text Display = [Position:(3,3)],[Item:1],[Size:100],[Color:Green]
20     WAIT WHILE ( SMART: Text to Speech (TTS) != 0 )
21     High-resolution Timer = 2.000sec
22     WAIT WHILE ( High-resolution Timer > 0.000sec )
23
24     ENDLESS LOOP
25     {
26        // Start Speech Recognition Process
27        SMART: Speech Recognition = Start (1)
28        // Wait for SR to complete
29        WAIT WHILE ( SMART: Speech Recognition != 0 )
30        VoiceCommand = SMART: Result of Speech Recognition
31        IF ( VoiceCommand != -1 )
32        {
33           CALL ProcessCommand
34           SMART: Result of Speech Recognition = 0
35        }
36        IF ( InvalidCommand == TRUE (1) )
37        {
38           InvalidCommand = FALSE (0)
39           JUMP Whistle
40        }
41     }
42  }
```

Fig. 3.46 Main Algorithm used in program "RC-UsingSpeech.tskx".

Fig. 3.46 describes the Main Algorithm which uses nested ENDLESS LOOPs:

1) The Outer ENDLESS LOOP is entered at Line 13. The "Whistle" label (line 15) is an important "Jump To" location that will be explained later with the Inner ENDLESS LOOP.

2) Line 16 shows that the user needs to whistle again to make the mobile device clear its display (line 17) and next voice out and display Text Item 1 (i.e. "ready") via lines 17-20.

3) Lines 21-22 shows that the robot would wait 2 seconds for the user to get ready to verbalize his or her command.

4) The Inner ENDLESS LOOP is entered at line 24 and the Voice Recognition task gets started by line 27 followed by a WAIT WHILE for this process to finish (line 29). When line 27 is executed the mobile device displays "Speak to Me" (see Fig. 3.47) and waits for the user to verbalize the command (within a given time-out period). When the user is finished with verbalizing and while line 29 is being executed, it would display "Recognizing" (see Fig. 3.48).

Fig. 3.47 "Speak to Me" display in program "RC-UsingSpeech.tskx".

Fig. 3.48 "Recognizing" display in program "RC-UsingSpeech.tskx".

5) Next, the Result of Speech Recognition is saved in Parameter "VoiceCommand" (line 30). It is a number, between 1 and 10, corresponding to the index of the "recognized" Text Item as shown in Fig. 3.43.

6) The first IF construct (lines 31-35) checks to see if the value saved in "VoiceCommand" is valid (i.e. different from "-1"). If it is valid, the Function "ProcessCommand" is called (line 33) and when this function is done, the parameter Result of Speech Recognition is reset to zero (line 34).

7) Function "ProcessCommand" (see Figs. 3.49 and 3.50) starts by checking if "VoiceCommand" is non-zero, i.e. an actual match is made with one of the Text/Voice Recognition Items shown in Fig. 3.43.

8) Next, the IF-ELSE-IF construct (Statements 107 to 131) is used to trigger the appropriate robot maneuvers (Forward, Backward, Turn Right, Turn Left or Stop) depending on the value found in "VoiceCommand" at runtime. Similarly, "VoiceCommand" is used to figure out the correct Slow and Fast Speed settings to use (lines 132 to 143 in Fig. 3.50).

```
FUNCTION ProcessCommand
{
    IF ( VoiceCommand != 0 )
    {
        IF ( VoiceCommand == Text Item 2 )
        {
            CALL Forward
            SMART: Text Display = [Position:(3,3)],[Item:2],[Size:100],[Color:White]
        }
        ELSE IF ( VoiceCommand == Text Item 3 )
        {
            CALL Backward
            SMART: Text Display = [Position:(3,3)],[Item:3],[Size:100],[Color:White]
        }
        ELSE IF ( VoiceCommand == Text Item 4 )
        {
            CALL TurnRight
            SMART: Text Display = [Position:(3,3)],[Item:4],[Size:100],[Color:White]
        }
        ELSE IF ( VoiceCommand == Text Item 5 )
        {
            CALL TurnLeft
            SMART: Text Display = [Position:(3,3)],[Item:5],[Size:100],[Color:White]
        }
        ELSE IF ( VoiceCommand == Text Item 6 )
        {
            CALL Stop
            SMART: Text Display = [Position:(3,3)],[Item:6],[Size:100],[Color:White]
        }
```

Fig. 3.49 Part 1 of Function "ProcessCommand" in program "RC-UsingSpeech.tskx".

```
132         ELSE IF ( VoiceCommand == Text Item 9 )
133         {
134             Speed = 200
135             TurnSpeed = 160
136             SMART: Text Display = [Position:(3,3)],[Item:9],[Size:100],[Color:White]
137         }
138         ELSE IF ( VoiceCommand == Text Item 10 )
139         {
140             Speed = 300
141             TurnSpeed = 200
142             SMART: Text Display = [Position:(3,3)],[Item:10],[Size:100],[Color:White]
143         }
144     }
145     ELSE IF ( VoiceCommand == 0 )
146     {
147         CALL Stop
148         InvalidCommand = TRUE (1)
149         SMART: Text Display = 0
150         SMART: Text to Speech (TTS) = Text Item 7
151         SMART: Text Display = [Position:(3,3)],[Item:7],[Size:100],[Color:Red]
152         WAIT WHILE ( SMART: Text to Speech (TTS) != 0 )
153         CALL Menu
```

Fig. 3.50 Part 2 of Function "ProcessCommand" in program "RC-UsingSpeech.tskx".

9) If no match is found for the user's verbal command, "VoiceCommand" would have a zero value and this "Invalid Command" case is dealt with by the construct comprising lines 145 to 153 (see Fig. 3.50). The robot is then stopped (Statement 147) and "InvalidCommand" is set to TRUE (line 148). Next, a Display and Text-to-Speech procedure is performed with Text Item 7 (i.e. "Invalid Command" – lines 149-152). Lastly, the Menu of Fig. 3.45 is redisplayed (line 153) and the program flow control returns to the main program at line 34 which goes on to line 36 (see Fig. 3.46).

10) Next, if "InvalidCommand" happened to be set to TRUE from inside the Function "ProcessCommand", then "InvalidCommand" was reset to FALSE (line 38), and the program flow control JUMPs (line 39) to LABEL "Whistle" (line 15) to restart the whole algorithm all over again.

Video 3.24 shows the runtime performance of the program "RC-UsingSpeech.tskx", wherein we could see that the Voice Recognition process required quite some time to finish, so the actual action followed about 3-5 seconds after the verbal command was issued. Although Speech Recognition is a "cool" technology, it is very clear that Voice Control cannot provide "responsive" maneuvers for our CarBot, thus Voice Control is not suitable for low-level control applications such as for motors.

3.6.5 Remote Control using Gradient Tool (Tilt Sensor) on Mobile Device

In this sub-section, we'll use the Gradient Tool provided by the R+m.PLAY700 App (see Fig. 2.3, under the "Sensor" group). This Gradient Tool uses the accelerometer and magnetometer sensors usually built-in on most mobile devices, however some entry-level devices may not be equipped with such sensors, thus this tool may not work properly for some mobile devices that the reader may use. It is recommended that readers check this functionality on their mobile devices first (see Video 3.25). Once the Gradient Tool (Tilt Sensor) is accessed via the "Editing" of a SMART Project within the R+m.PLAY700 App, the mobile device would display a screen resembling the one shown in Fig. 3.51. In this instance, the mobile device was tilted such that its Front Edge was UP (18 degrees) and its Right Edge was also UP (28 degrees), within respect to the horizontal plane. Readers should notice that "negative" gradients are not "reported" by this tool.

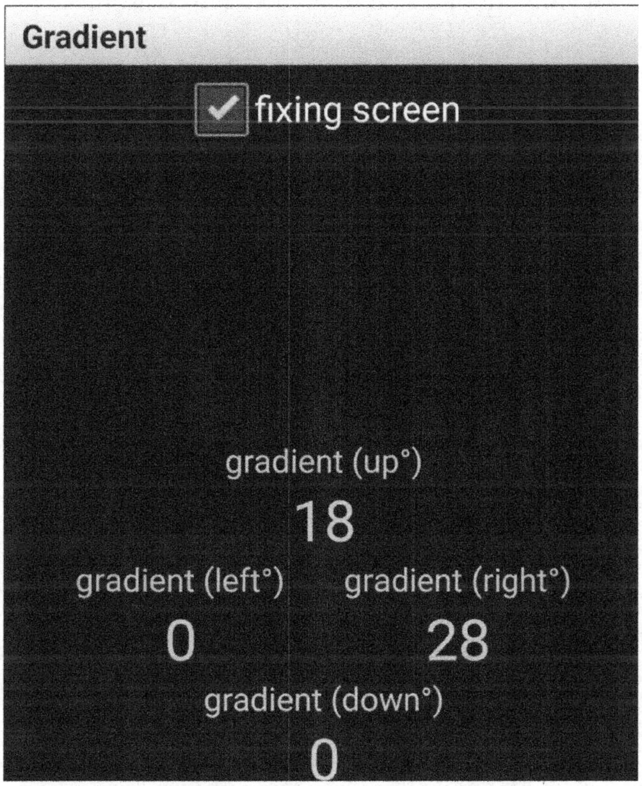

Fig. 3.51 Gradient Tool (Tilt Sensor) in action.

For this application "RC-Tilt-Sensor.tskx", the **Tilt Sensor is used as a joy-stick controller**, thus the nomenclature of the sensing parameters and their use in various conditional statements would read, at places, **like a complete reversal of the logic used with the RC-100 controller!**

Fig. 3.52 describes the Initial Actions:

Play Melody 2 (lines 5-7).

Set Forward/Backward Speed to 256 (line 8).

Set Left/Right Speed to 160 (line 9).

Set DelayTime to 250 ms (line 10).

```
// Init Action
Buzzer Timer = Melody Time
Buzzer Index = Melody No.2 (2)
WAIT WHILE ( Buzzer Timer > 0.0sec )
SpeedFB = 256
SpeedLR = 160
DelayTime = 250
```

Fig. 3.52 Initial Actions in “RC-Tilt-Sensor.tskx”.

```
11 ENDLESS LOOP
12 {
13   SMART: Screen Rotation = Landscape Mode (2)
14   Gradient_Up = SMART: Acceleration Sensor (U)
15   Gradient_Down = SMART: Acceleration Sensor (D)
16   Gradient_Left = SMART: Acceleration Sensor (L)
17   Gradient_Right = SMART: Acceleration Sensor (R)
18   SMART: Number Display = [Position:(3,2)],[Item:Gradient_Up],[Size:100],[Color:White]
19   SMART: Number Display = [Position:(2,3)],[Item:Gradient_Left],[Size:100],[Color:White]
20   SMART: Number Display = [Position:(4,3)],[Item:Gradient_Right],[Size:100],[Color:White]
21   SMART: Number Display = [Position:(3,4)],[Item:Gradient_Down],[Size:100],[Color:White]
22   IF ( Gradient_Up < 5 && Gradient_Down < 5 && Gradient_Left < 5 && Gradient_Right < 5 )
23   {
24     CALL Stop
25   }
```

Fig. 3.53 Part 1 of Main Algorithm for “RC-Tilt-Sensor.tskx”.

Fig. 3.53 shows Part 1 of the Main Algorithm (with the familiar EndLess Loop) whereas:

1) This Endless Loop starts by constraining the mobile device into a landscape display mode (line 13). This action is needed to prevent the mobile device from switching, unintentionally, between its two display modes (portrait and landscape) later during runtime of this program when this device is manipulated as a joy-stick, as this would confuse the user considerably. This is a Human-Machine-Interface issue more than a Robotics one. The reader is encouraged to comment out Line 13 to see its effects at runtime.

2) The mobile device’s reference plane is the horizontal plane. Thus all 4 “Gradients or Accelerations” parameters (in degrees) would be reported as 0 degree when the device is positioned flat in a horizontal manner. Furthermore, the device would only provide POSITIVE Gradient values whenever they were

ABOVE the horizontal plane. For example, if the Front edge of the phone is above the horizon, the Gradient_Up parameter would show a positive value while the Gradient_Down parameter would show a "zero" value (although logically it should have been a "negative" value and equal in magnitude to Gradient_Up). Lines 14-17 record the U-D-L-R gradients and lines 18-21 display them on the mobile display in a cross configuration centered on the screen.

3) The author wants the robot to be still when the mobile device is held "mostly" in a horizontal plane thus the use of the "5-degrees" as strict upper limits in line 22 of Fig. 3.53.

```
26 ELSE
27 {
28     IF ( Gradient_Down >= 5 )
29     {
30         CALL Forward
31         CALL Delay
32     }
33     IF ( Gradient_Up >= 5 )
34     {
35         CALL Backward
36         CALL Delay
37     }
38     IF ( Gradient_Right >= 5 )
39     {
40         CALL LeftTurn
41         CALL Delay
42     }
43     IF ( Gradient_Left >= 5 )
44     {
45         CALL RightTurn
46         CALL Delay
47     }
48 }
```

Fig. 3.54 Part 2 of Main Algorithm for "RC-Tilt-Sensor.tskx".

Fig. 3.54 describes Part 2 of the Main Algorithm whereas:

1) When any of the Gradients U-D-L-R are greater or equal to 5 degrees, i.e. when line 22 evaluates to FALSE, the ELSE branch (line 26 in Fig. 3.54) would be taken by the algorithm.

2) Next, four parallel IFs constructs (lines 28 to 47) are used in this ELSE branch as the author wants the robot to be able to move in more than one direction (U-D-L-R) at any one time to achieve smoother curved trajectories.

Video 3.26 shows the runtime performance of this program "RC-Tilt-Sensor.tskx".

3.7 Autonomous Line Tracking & Obstacle Avoidance with NIR Sensors

For this Section 3.7, we will be using a CarBot configuration as described in Fig. 3.55. The goal is to program this CarBot to go forward and follow an arbitrary track created using standard black electrical tape laid out on brown Craft paper. This CarBot would have to use its Left and Right NIR sensors to keep it "on-track". During its progress along this track, if the CarBot finds a frontal obstacle, it should be able to swing around, re-discover the black track and follow it again but in the opposite direction.

Fig. 3.55 CarBot used as Autonomous Line Tracker & Obstacle Avoider.

Let's first analyze how the three NIR Sensors behave when the CarBot is laid on the black track under the four main conditions that would occur in this challenge: a) when the track is under the bot's Left IR

Sensor; b) when the bot is centered on the track; c) when the track is under the bot's Right IR Sensor; and d) when the bot is gone completely off-track (see Fig. 3.56).

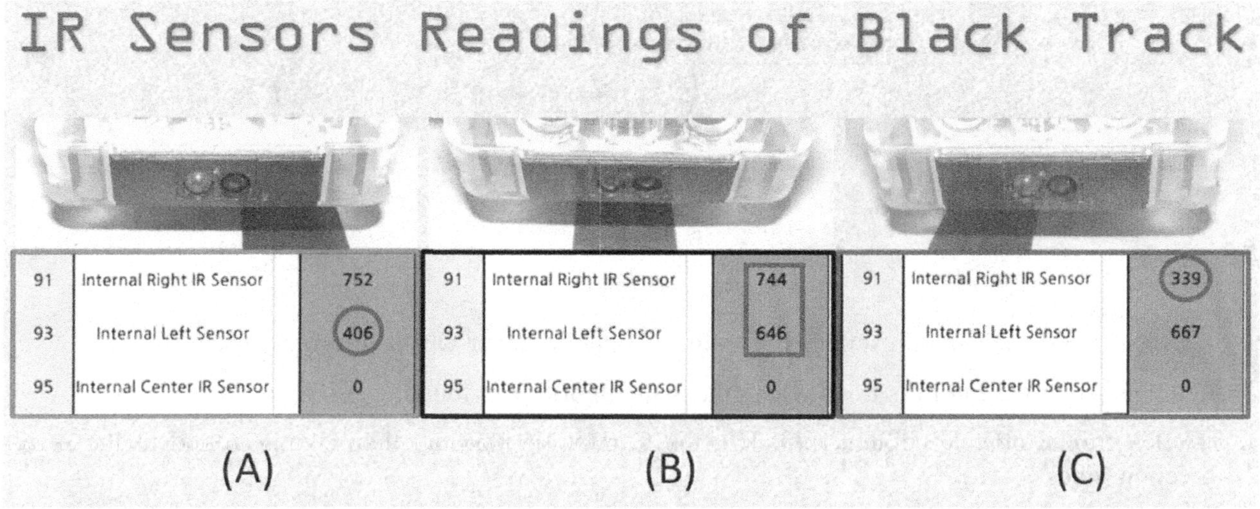

Fig. 3.56 Values for L-R-C IR Sensors for 3 configurations between CarBot and Black Track.

In Fig. 3.56 and going from Column (A) to Column (C) to check for the readings from the Right IR Sensor (Address 91): ones can see that its value is around 740-750 whenever it sees the background Craft paper, and that its value is reduced to around 340 when this sensor is "seeing" the black track. Similarly, for the Left IR Sensor (Address 93), it reads around 400 when it "sees" the black track and between 640 and 670 when it "sees" the Craft paper surface. Most importantly, Column (B) shows that the spacing between the Left and Right IR Sensors is larger than the width of the black electrical tape as both sensors read "high" here (e.g. 744 and 646) - in other words, both sensors are only seeing the brown paper. **Thus, it is physically impossible for this CarBot to figure out via its Left and Right NIR sensors whether it is centered on the black track (case b) or that it is completely off-track (case d)!**

The previous experimental analysis shows that there are only two conditions (mutually exclusive) that can be monitored unequivocally by this CarBot hardware:

1) If the Left IR Sensor can see the black track then it should turn left in its attempt to re-align the bot with the black tract.

2) If the Right IR Sensor can see the black track then it should turn right in its attempt to re-align the bot with the black tract.

Otherwise, the robot really cannot be "sure" of anything else, and so it should continue its default maneuver (i.e. going forward). From previous sections, we are then expecting to see an IF-ELSE-IF structure for three mutually exclusive alternatives in the programming solution for this challenge.

Fig. 3.57 shows the Initial Actions for the solution program "LineFollower-ObstacleAvoider.tskx", whereas the CM-50 plays Melody 2 (lines 5-7) then sets initial values for 4 Parameters - Speed, TimeDelay, LineDistance and ObjectDistance (lines 8-11). The "365" value assigned to Parameter LineDistance is just a number between the values "339" and "406" that were reported for both IR Sensors (see Fig. 3.56 and blue circles in columns (A) and (C)).

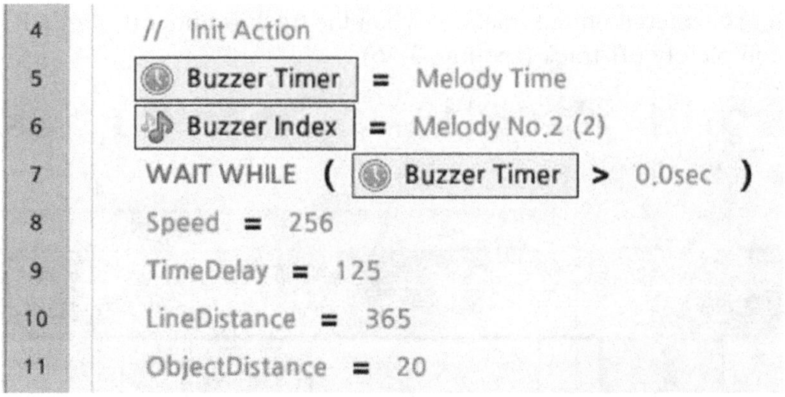

Fig. 3.57 Initial Actions for program “LineFollower-ObstacleAvoider.tskx”.

Fig. 3.58 shows the EndLess Loop where the main algorithm is found:

1) If a frontal object is found (line 16), the CarBot should stop then swing around to the reverse direction (lines 18-19).

2) Otherwise, the programming logic flows into an IF-ELSE-IF structure whereas:

If only the Left IR Sensor “sees” black (line 24) then the “LeftTurn” Function is called (line 26). Otherwise, the Right IR Sensor is checked next (line 29), and if only this Right Sensor “sees” black then Function “RightTurn” is called (line 32).

For all other situations (i.e. the ELSE on line 34), Function “Forward” is executed (line 37).

```
ENDLESS LOOP
{
    // If there is an obstacle in front (using IR Center)
    IF ( IR Center > ObjectDistance )
    {
        CALL Stop
        CALL SwingAround
    }
    ELSE
        // No obstacle in front of CarBot, so OK to follow black track
    {
        IF ( IR Left < LineDistance && IR Right >= LineDistance )
        {
            // Turns left when ONLY the left sensor detects black.
            CALL LeftTurn
        }
        ELSE IF ( IR Left >= LineDistance && IR Right < LineDistance )
        {
            // Turns right when ONLY the right sensor detects black
            CALL RightTurn
        }
        ELSE
        {
            // Moves forward for all other cases.
            CALL Forward
        }
    }
}
```

Fig. 3.58 Main Algorithm for program "LineFollower-ObstacleAvoider.tskx".

The Function "SwingAround" (Fig. 3.59) describes a combination of "timed" and "conditional" right turns to accommodate for the fact that the "obstacle" can be found at any spot on the black track:

First, a 0.125s timed right-turn is used (lines 82-85) because at this point in time, the CarBot is mostly lined up with the black track and it just needs to be nudged a bit to the right to make sure that Right IR Sensor is now "seeing" the Craft paper (i.e. "white" - the condition (IR_Right > LineDistance) is TRUE).

```
79 FUNCTION SwingAround
80 {
81     // Turn right for 0.125 second to clear IR-Right sensor off the track
82     PORT[1]:Geared Motor = CCW:0 (0.00%) + 250
83     PORT[2]:Geared Motor = CCW:0 (0.00%) + 250
84     High-resolution Timer = 0.125sec
85     WAIT WHILE ( High-resolution Timer > 0.000sec )
86 
87     // Keep on turning right as long as IR-Right sees white
88     WAIT WHILE ( IR Right > LineDistance )
89 
90     // IR-Right now sees Black, but OK to keep on turning right until IR-Right sees White again
91     WAIT WHILE ( IR Right <= LineDistance )
92     // Bot should be lining up with the track but in the reverse direction at this point
93 }
```

Fig. 3.59 Function "SwingAround" for program "LineFollower_ObstacleAvoider.tskx".

Next, the reader needs to recall that the previous commands for the CarBot to turn right (lines 82-83) are still in effect. The 2 WHILE loops (lines 88 and 91) are "conditional" loops so that the CarBot can adjust to "any" section of the arbitrary track where the "SwingAround" maneuver may be needed. These 2 loops are used to ensure that an orderly process is followed to detect whether the Right IR Sensor is seeing "black" or "white" (e.g. brown Craft paper):

The first loop (line 88) is active whenever Right IR sees "white", i.e. whenever the CarBot has **not** yet crossed back into the black track. When this situation happens, i.e. when IR Right sees "black", the first loop's conditional expression becomes false and this loop is exited, and the second loop is entered (line 91).

The purpose for the second loop (line 91) is to keep the CarBot turning right some more until IR Right sees "white" (i.e. it has just passed the other edge of the black track). At this point in time, the CarBot should be mostly lined up with the black track (but in the reverse direction), thus this Function is exited and another iteration of the Main Endless Loop is executed, resulting in the Carbot's default maneuver of Going Forward.

Video 3.27 shows the performance of this algorithm on different shapes of a closed loop black track. Ones can see that this algorithm cannot handle "sharp" turns, as it would get off-track at these spots.

3.8 Using Mobile Video Camera for Color Line Tracking

In this section, we'll use the Line Detection tool provided by the R+m.PLAY700 App (see Fig. 2.3, under the "Vision" group). This tool allows some rudimentary "real-time" image processing functions on the mobile device that are however accessible to the CM-50 via TASK programming. Fig. 3.60 displays the bottom view of a CarBot with a SmartPhone mounted on top, showing the camera view port and the built-in LED light source that can be used to "light up" the viewed scene.

Fig. 3.60 Bottom View of a CarBot equipped with a SmartPhone.

The R+m.PLAY700 App divides the phone camera's entire sensing area into 25 equal zones, i.e. 5 rows by 5 columns, just like how it partitions its 25 Display Areas for Texts, Shapes and Numbers, and also for its 25 Touch Areas (see Fig. 3.34). Video 3.28 shows how to check the functionality of this tool using a piece of white paper with different color lines on it. Fig. 3.61 is a screen capture of the mobile device while it is supposed to track a Red Line. In this figure, the reader can see that this tool classifies a Yellow Line erroneously as a second Red Line (see the little window in Upper Left corner of Fig. 3.61) – this is the limitation of the current R+m.PLAY700 App.

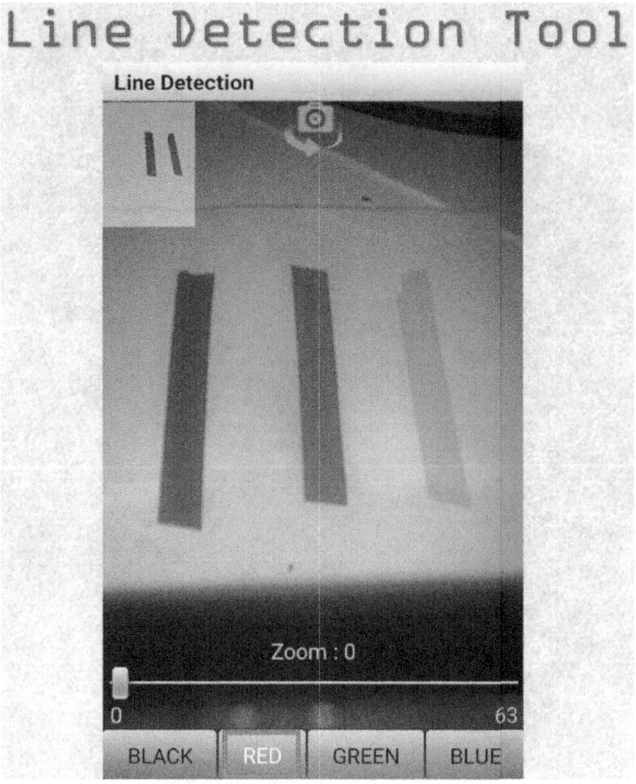

Fig. 3.61 Performance of Line Detection Tool for Red and Yellow lines.

The current Line Detection tool uses only the top row of the sensing zones (i.e. zones 1-5), mainly because of the limitation in throughput for the Bluetooth communication channel which would be overwhelmed when data from all 25 zones were streamed back to the CM-50 robot to process. When the CarBot as configured in Fig. 3.61 is laid on top of a Red line, Fig. 3.62 shows two important features to be considered when using this tool inside a TASK program:

1) The image captured is out of focus because the Red line turns out to be set too close to the camera, but this will have no effect in this particular application where we only care to know about which zone, among zones 1 through 5, contains the Red line (i.e. a fuzzy red blob in left image in Fig. 3.62).

2) "Zoom" may have to be set so that the top row of the camera sensor (i.e. zones 1 to 5) can detect enough of the fuzzy Red line so that it can display a specific zone number (which is "3" in Fig. 3.62) where it has found a "red blob". "Zoom" level is set to "4" in Fig. 3.62. Please note that this "issue" is due to the LED lighting condition used by the author, i.e. a bit too dark at the top edge of the camera sensor in left image and zooming takes care of this problem as shown in right image. If the reader uses ambient diffuse light, this issue may not arise for the reader at all – so be sure to check your actual usage situation to determine whether "zooming" is needed or not (see Fig. 3.63 where zoom was not needed and the detected zone was found to be 2).

Fig. 3.62 Tracking a Red line without zoom (left) and with zoom (right).

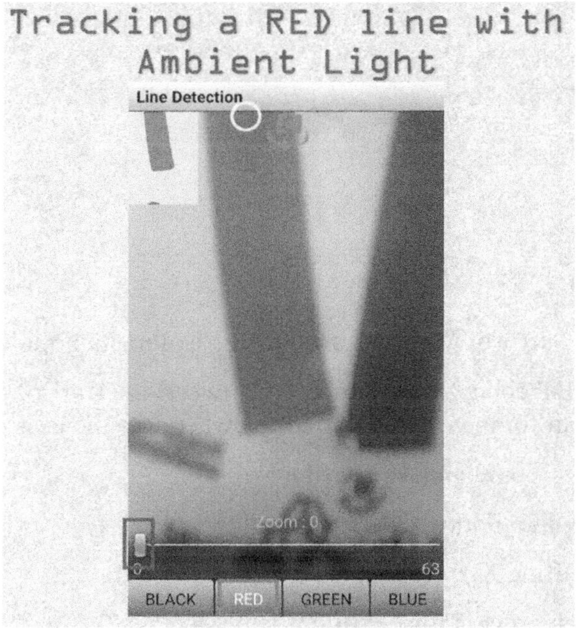

Fig. 3.63 Tracking a Red line with Ambient Light.

The Color Line Detection process has 3 general steps:

1) Set SMART DEVICE "Camera Sensor" to "Line Detection Mode".

2) Set SMART DEVICE "Vision – Tracking Color" to a chosen color among 4 pre-defined "Colors": Black (=2), Red (=3), Green (=4) and Blue (=5).

3) Read SMART DEVICE "Vision – Line Detection Area" to obtain a numerical value which would be between 1 and 5 when the "assigned" color is found, otherwise it would return a zero value.

The solution TASK program to this challenge is called "ColorLineFollower.tskx" and it uses Nested Loops: i.e. an external ENDLESS LOOP along with an internal LOOP WHILE (depending on the detected zone number). Fig. 3.64 shows Part 1 of the main ENDLESS LOOP:

```
6   ENDLESS LOOP
7   {
8       // User needs to make sure that robot is connected to mobile device. User claps ha
9       Result of Sound Counter = 0
10      WAIT WHILE ( Result of Sound Counter == 0 )
11      SMART: Screen Rotation = Portrait Mode (1)
12      SMART: Text Display = 0
13      SMART: Text to Speech (TTS) = Text Item 1
14      SMART: Text Display = [Position:(3,3)],[Item:1],[Size:100],[Color:Green]
15      SMART: Camera Selection = Back Camera (1)
16      SMART: Camera Zoom = 0
17      WAIT WHILE ( SMART: Text to Speech (TTS) != 0 )
18      // Setting Camera Sensor to Line Detection mode
19      SMART: Camera Sensor = Line Detection Mode (4)
20      SMART: Text Display = 0
21      // User sets LineColor to track: 2 for Black; 3 for Red; 4 for Green; 5 for Blue
22      LineColor = Red Line (3)
23      SMART: Tracking Color (Line-Tracer) = LineColor
```

Fig. 3.64 Part 1 of Main Algorithm for "ColorLineFollower.tskx".

1) To make sure that all BT connections are made between the CarBot and the mobile device, lines 9 and 10 work together to wait for the user to clap hands whenever the user is ready.

2) Next, a series of SMART commands are used to:

Put the Mobile Display into Portrait Mode (line 11).

Clear the Text Display (line 12).

Start the Text-to-Speech process for Text Item 1 (e.g. "ready") (line 13).

Display Text Item 1 ("ready") in the display's center using Green letters (line 14).

Use the Back Camera (line 15).

Set the camera's zoom level to zero (line 16).

Wait for the Text-to-Speech process to finish (line 17).

3) Line 19 sets the Camera Sensor into the Line Detection Mode and line 20 clears the screen the second time, getting ready for displaying debugging information from Part 2 of this algorithm. At this step the Camera Sensor is not yet told of what Color to detect.

4) Parameter "LineColor" (line 22) must be set by the user to designate the Color Type to be detected by the Camera Sensor: use "2" for Black, "3" for Red, "4" for Green and "5" for Blue. In this example, the author uses "3" for "Red".

5) Parameter "LineColor" ("Red") is then assigned to the "Tracking Color – Line Tracer" Parameter (line 23). This tells the Camera Sensor to look for any potential "Red Line" from now on. It is important to notice that lines 19, 22 and 23 are designed to work together in this order.

Part 2 of this algorithm starts with line 25 in Fig. 3.65:

1) Line 25 triggers the "Line Detection Area" procedure to run, that is for the Camera Sensor to search among the display zones 1 through 5 for the presence of any group of Red pixels in each of these zones. The zone with the biggest number of Red pixels "wins" and its zone number [1-5] is retrieved from this procedure and assigned to Parameter LinePosition. If no Red pixel was found among all five zones, then a zero value is returned and assigned to Parameter LinePosition.

2) Next is the crucial internal LOOP WHILE structure (lines 26-43) which uses the Conditional Statement (LinePosition != 0). If LinePosition has a value between 1 and 5 which means that a group of Red pixels had been found among one of the 5 zones tested, then this Condition evaluates to TRUE and this LOOP WHILE is entered at line 27. If LinePosition has a zero value, which means that no Red pixel had been found among the 5 zones, then this LOOP WHILE is exited and line 44 is executed next.

```
24      // Obtain Initial LinePosition, i.e. zone number where LineColor is detected
25      LinePosition = [ SMART: Line Detection Area ]
26      LOOP WHILE ( LinePosition != 0 )
27      {
28          CALL DisplayColorPosition
29          IF ( LinePosition == 3 )
30          {
31              CALL Forward
32          }
33          ELSE IF ( LinePosition == 1 || LinePosition == 2 )
34          {
35              CALL RotateLeft
36          }
37          ELSE IF ( LinePosition == 5 || LinePosition == 4 )
38          {
39              CALL RotateRight
40          }
41          // Obtain latest LinePosition, i.e. zone number where LineColor is detected
42          LinePosition = [ SMART: Line Detection Area ]
43      }
44      CALL Stop
45      [ SMART: Text to Speech (TTS) ] = Text Item 2
46      [ SMART: Text Display ] = [Position:(3,1)],[Item:2],[Size:125],[Color:White]
47      WAIT WHILE ( [ SMART: Text to Speech (TTS) ] != 0 )
48  }
```

Fig. 3.65 Part 2 of Main Algorithm for "ColorLineFollower.tskx".

3) If LinePosition is non-zero, the LOOP WHILE is executed, first by calling Function "DisplayColorPosition" (line 28) which displays Parameters LineColor and LinePosition on the same line (for debugging purposes, the reader can comment out line 28 if needed). Next an IF-ELSE-IF structure is used to decide whether the CarBot should Go Forward (when LinePosition equals 3), or that it should Rotate Left (when LinePosition equals 1 or 2), or that it should Rotate Right (when LinePosition equals 4 or 5), all designed for the CarBot to track the colored line. The last action for this LOOP is to "refresh" the value of LinePosition with another access to the "Line Detection Area" procedure (line 42). Then the LOOP WHILE structure is ready again for another iteration if LinePosition turns out to be non-zero, otherwise the program flow control is transferred to line 44 which stops the CarBot.

4) Lines 45-47 voices out and displays Text Item 2 which is “No Color Found”, then the program flow control is transferred to the outer ENDLESS LOOP (i.e. line 7), meaning that CarBot is now waiting on the user to clap hands if the user wants the CarBot to re-start its task of following a Red Line.

Video 3.29 shows the runtime performance of this program using ambient light.

3.9 Multimedia Moody Dog with R+m.PLAY700

This project illustrates the use of graphics display and audio files on the mobile device and shows how to integrate their timing with sensors on board the CM-50.

The corresponding TASK code is called “MoodyDogBot.tskx”. It uses two Audio files: “Small-dog-barking.mp3” and “Angry-dog.mp3”. It uses 4 background images: “DogBot.png”, “MoodyFella.png”, “FriendlyDog.png” and “AngryDog.png”. The reader will need to create a Custom Project inside the R+m.PLAY700 App and these 6 files need to be stored in specific folders of this “Moody Dog” SMART project (see Figs. 3.66 and 3.67). Video 3.30 demonstrates how to set up this project on a mobile device).

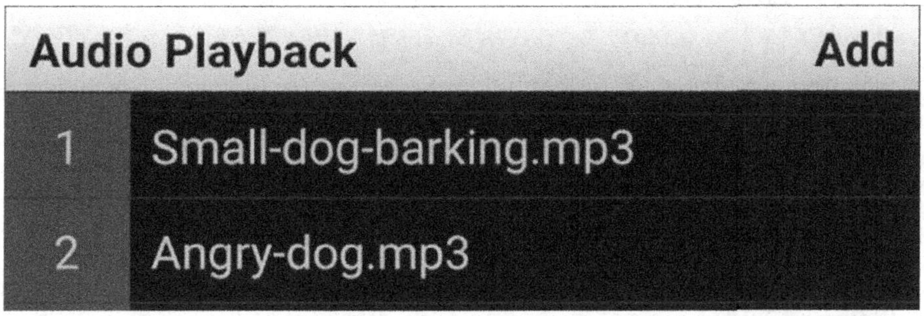

Fig. 3.66 Audio Files used by Moody Dog SMART Project.

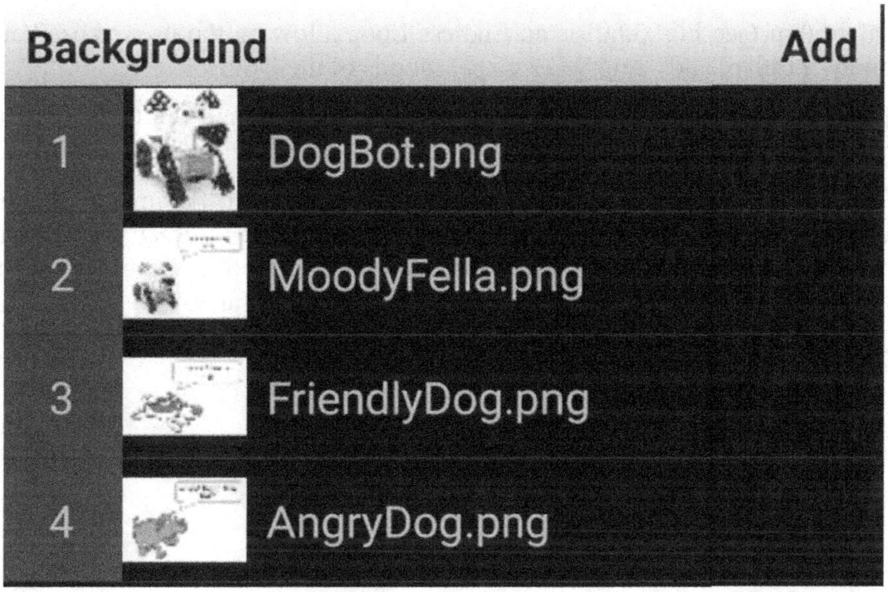

Fig. 3.67 Background Graphics Files used by Moody Dog SMART Project.

Fig 3.68 shows the Initial Actions (Part 1) of the program "MoodyDogBot.tskx":

First, the CM-50 plays Melody 2 (lines 5-7) and waits for the user to whistle into the mobile device (line 9).

```
4   // Initial Actions
5   Buzzer Timer = Melody Time
6   Buzzer Index = Melody No.2 (2)
7   WAIT WHILE ( Buzzer Timer > 0.0sec )
8   // Waiting for user to whistle into mobile device when ready, then displays Background Image 1
9   WAIT WHILE ( SMART: Noise (dB) < 50 )
10  SMART: Background Image = 1
11  // Announce and Display "Ready!"
12  SMART: Text to Speech (TTS) = Text Item 1
13  SMART: Text Display = [Position:(3,3)],[Item:1],[Size:200],[Color:Green]
14  WAIT WHILE ( SMART: Text to Speech (TTS) != 0 )
15  High-resolution Timer = 2.000sec
16  WAIT WHILE ( High-resolution Timer > 0.0sec )
17  SMART: Text Display = 0
```

Fig. 3.68 Part 1: Initial Actions in "MoodyDogBot.tskx"

Next, the mobile device displays Background Image 1 (line 10). It also announces and displays Text Item 1 "Ready!" (lines 12-14).

The CM-50 then waits for 2 seconds (lines 15-16), and sends a Text Display command telling the mobile device to clear its Text Screen (line 71) to get ready for Part 2 of the algorithm.

Part 2 of the algorithm (see Fig. 3.69) is an Endless Loop allowing the user to use hand claps to select between the "Friendly" (1 clap) and "Angry" (2 claps) moods of the DogBot:

The Background Image 2 "Moody Fella~" is displayed (line 22), and the CM-50 waits for the user to clap once for the "Friendly Dog" mood and twice for the "Angry Dog" mood (line 23-25).

Next, depending on the number of claps recorded (1 or 2, saved in Parameter Claps), one of the two parallel IFs constructs would be triggered (lines 27-30 and 32-35), effectively CALLing upon either Function "Friendly" or Function "Angry". Both Functions share the same code design. After these functions are executed, a new iteration of the EndLess Loop begins anew.

```
ENDLESS LOOP
{
    //  Moody Fella~ section
    SMART: Background Image = 2
    Result of Sound Counter = 0
    WAIT WHILE ( Result of Sound Counter == 0 )
    Claps = Result of Sound Counter
    //  Friendly section
    IF ( Claps == 1 )
    {
        CALL Friendly
    }
    //  Angry section
    IF ( Claps == 2 )
    {
        CALL Angry
    }
}
```

Fig. 3.69 Part 2: MAIN LOOP of "MoodyDogBot.tskx"

```
FUNCTION Friendly
{
    SMART: Background Image = 3
    WAIT WHILE ( IR Center < 20 )
    PORT[1]:Geared Motor = CW:0 (0.00%) + 500
    PORT[2]:Geared Motor = CCW:0 (0.00%) + 500
    SMART: Play Audio 1 = 1
    High-resolution Timer = 0.500sec
    WAIT WHILE ( High-resolution Timer > 0.000sec )
    PORT[1]:Geared Motor = CW:0 (0.00%) + 0
    PORT[2]:Geared Motor = CCW:0 (0.00%) + 0
    WAIT WHILE ( SMART: Play Audio 1 != Not Used (0) )
}
```

Fig. 3.70 Function "Friendly" of "MoodyDogBot.tskx"

Fig. 3.70 describes how the Function "Friendly" is constructed:

Background Image 3 "FriendlyDog.png" is displayed on the mobile device (line 3) and the CM-50 (i.e. DogBot) watches its Center IR Sensor Port to see if any object is appearing there and close enough to make its reading greater than 20 (line 42).

If there is an object closing in on the DogBot, line 42 is exited and the CM-50 turns on its two motors at Speed 500 (i.e. the DogBot runs forward towards the object detected).

Next, Audio Channel 1 on the mobile device is used to start playing Audio File 1 (i.e. "Small-dog-barking.mp3"), and the Hi-Res Timer is set to elapse 0.5 second (lines 46-47).

When the 0.5 s delay is done, power to the 2 motors are turned off, but this Play Audio Function is still active until the Audio Channel 1 is finished playing (line 50 – sorry for the "double negative", but it needs to be used here).

Video 3.31 shows the runtime performance of this SMART project.

3.10 Using Musical Instruments on CM-50 and Mobile Device

When using the CM-50 in concert with the R+m.PLAY700 running on a mobile device, there are many audio output devices that can be activated at the same time:

The Buzzer on the CM-50, providing musical scales and melodies (see Section 3.3.1).

Two simultaneous audio sources on the mobile device, capable of playing most audio files (see Section 3.9).

Musical scales and Octave levels with One Musical Instrument chosen among 128 possible Instruments (more details later in this Section 3.10).

First, let's examine the structure of the SMART CONSTANT that can be used to play a SMART Musical Instrument. This SMART CONSTANT has 4 bytes, but only the lower 3 bytes are used:

The first (lowest) byte is kept for MUSICAL SCALE or NOTE (possible values are 1 to 12): e.g. "1" for "Do", "2" for "Do#", etc… with "12" corresponding to "Shi".

The second byte is kept for OCTAVE (possible values are 1 through 10).

The third byte is kept for INSTRUMENT TYPE (128 possible types, supported on Android devices only): "1" for "Acoustic Grand Piano" and "128" to "Gunshot", with other instruments in between.

In this application, the goals are to make the CM-50's buzzer cycle through its Melody Index from 0 to 24, and to make the mobile device cycle through its Musical Instrument Type from 1 to 128. For the two "mobile" audio sources, the plan is to let each audio source play its own audio file continuously.

The program "MusicalInstruments.tskx" shows how to set up needed parameters and how to activate these 4 audio sources within an Endless Loop. Fig. 3.71 describes the Initial Actions:

The user needs to whistle into the mobile device to get this application started (line 6), next the mobile display is set to Portrait Mode and its Text Display cleared (lines 7-8).

Lines 9 to 19 initialize the 11 parameters used in this program. MelodyIndex loops from "0" to "24", and "InstrumentType" cycles from "1" through "128". Parameters "Octave", "DoNote", "ReNote" and

"MiNote" remain constant throughout this program. "InstrumentValueTemp1" and "InstrumentValueTemp2" are used to compute intermediary results that are later saved as final values into three Combined Instrument Parameters named "InstrumentValue1", "InstrumentValue2" and "InstrumentValue3". Combined Instrument Parameters have three types of information combined in them: Instrument Type [1-128], Octave Level (set to 3), and the musical notes Do [1], Re [3] and Mi [5].

```
WAIT WHILE ( SMART: Noise (dB) < 50 )
SMART: Screen Rotation = Portrait Mode (1)
SMART: Text Display = 0
MelodyIndex = 0
InstrumentType = 1
Octave = 3
DoNote = 1
ReNote = 3
MiNote = 5
InstrumentValue1 = 0
InstrumentValue2 = 0
InstrumentValue3 = 0
InstrumentValueTemp1 = 0
InstrumentValueTemp2 = 0
```

Fig. 3.71 Initial Actions in "MusicalInstruments.tskx".

Fig. 3.72 described Part 1 of the MAIN Endless Loop which controls of the CM-50's buzzer:

```
ENDLESS LOOP
{
    // Using Buzzer on Robot Controller - going through 25 melodies [0-24]
    SMART: Text Display = [Position:(2,1)],[Item:1],[Size:100],[Color:White]
    SMART: Number Display = [Position:(4,1)],[Item:MelodyIndex],[Size:100],[Color:White]
    Buzzer Timer = Melody Time
    Buzzer Index = MelodyIndex
    MelodyIndex = MelodyIndex + 1
    IF ( MelodyIndex == 25 )
    {
        MelodyIndex = 0
    }
```

Fig. 3.72 Part 1 of Endless Loop – Controlling the Buzzer.

Lines 24-25 displays on the mobile device the value of "MelodyIndex" which is about to be played by the buzzer. Text Item 1 is set to "Melody". Line 25 has a Custom SMART Constant for its RHS wherein Variable Parameter "MelodyIndex" is used in the "Item" field (see Fig. 3.73).

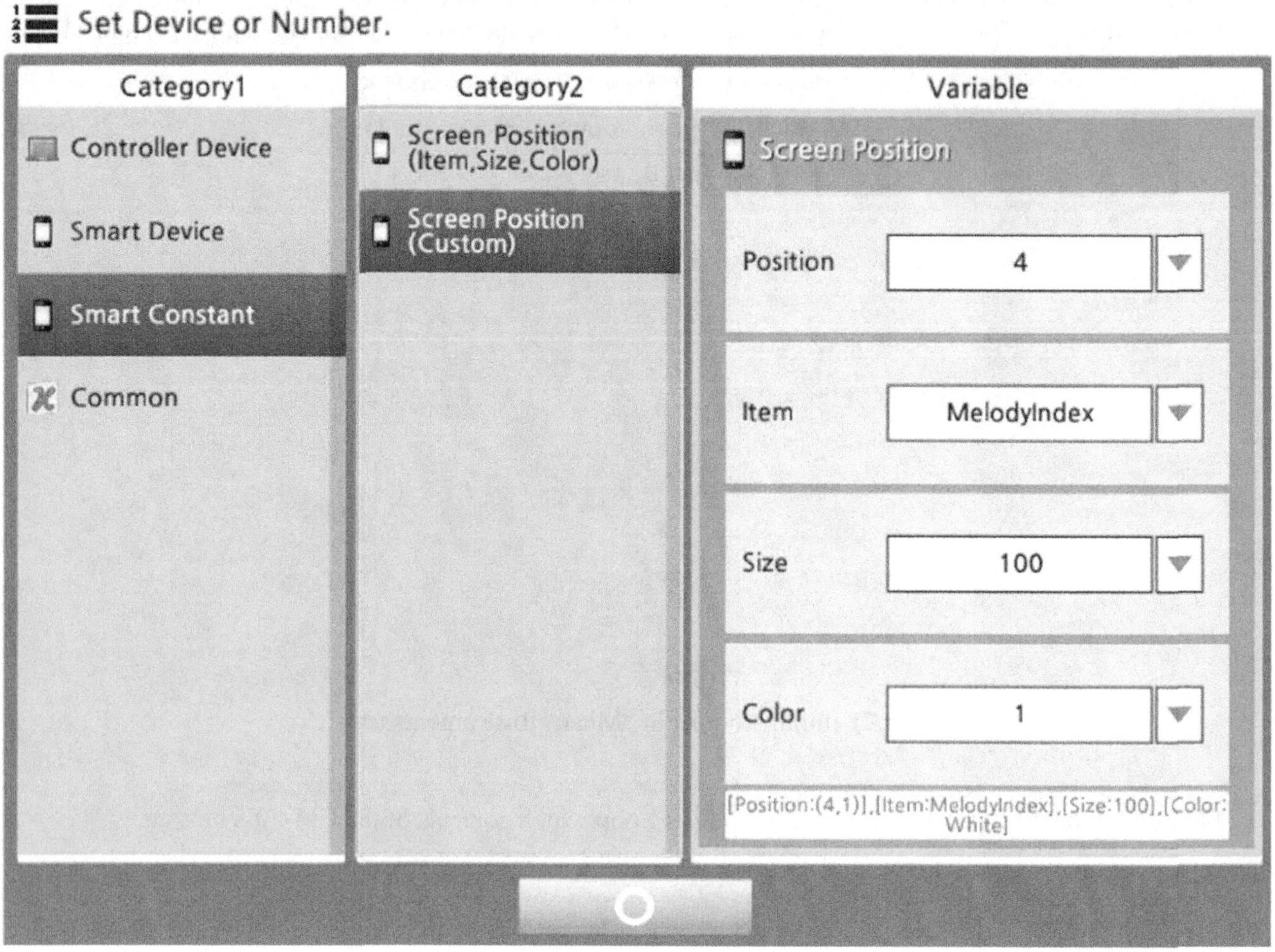

Fig. 3.73 Custom Constant used in RHS of Line 25.

Lines 26-27 activates the CM-50's buzzer using the current "MelodyIndex" value.

Lines 28-32 increments MelodyIndex by "1" for each iteration of the Endless Loop, however when it reaches "25", it is reset to "0".

Fig. 3.74 describes Part 2 of the MAIN Endless Loop which controls two simultaneous Audio Channels on the mobile device:

Statements 34 to 37 played Audio .Item 3 on Play Audio channel 1 continuously.

Statements 38 to 41 played Audio Item 4 on Play Audio channel 2 continuously.

```
// Using Audio Outputs 1 and 2 on Mobile Device playing Audio Items 3 & 4 continuously
IF ( SMART: Play Audio 1 == Not Used (0) )
{
    SMART: Play Audio 1 = Audio Item 3
}
IF ( SMART: Play Audio 2 == Not Used (0) )
{
    SMART: Play Audio 2 = Audio Item 4
}
```

Fig. 3.74 Part 2 of Endless Loop – Controlling two Audio Channels on Mobile Device.

Fig. 3.75 shows Part 3 of the main Endless Loop which controls the Musical Instrument on the mobile device:

```
// Using Musical Instrument on Mobile Device for 5 seconds - going through 128 instruments [1-128]
SMART: Text Display = [Position:(2,3)],[Item:2],[Size:100],[Color:Green]
SMART: Number Display = [Position:(4,3)],[Item:InstrumentType],[Size:100],[Color:Green]
InstrumentValueTemp1 = InstrumentType * 0000 0000 0000 0001 0000 0000 0000 0000
InstrumentValueTemp2 = Octave * 0000 0000 0000 0000 0000 0001 0000 0000
InstrumentValueTemp1 = InstrumentValueTemp1 + InstrumentValueTemp2
InstrumentValue1 = InstrumentValueTemp1 + DoNote
InstrumentValue2 = InstrumentValueTemp1 + ReNote
InstrumentValue3 = InstrumentValueTemp1 + MiNote
InstrumentType = InstrumentType + 1
IF ( InstrumentType == 129 )
{
    InstrumentType = 1
}
SMART: Play a musical instrument = InstrumentValue1
WAIT WHILE ( SMART: Play a musical instrument > 0 )
SMART: Play a musical instrument = InstrumentValue2
WAIT WHILE ( SMART: Play a musical instrument > 0 )
SMART: Play a musical instrument = InstrumentValue3
WAIT WHILE ( SMART: Play a musical instrument > 0 )
}
```

Fig. 3.75 Part 3 of Endless Loop – Controlling Musical Instruments on Mobile Device.

Lines 43-44 display the Instrument Type about to be played. Text Item 2 is set to "Instrument". Line 44 has another Custom SMART Constant similar to the one shown in Fig. 3.73, but for Parameter "InstrumentType".

Line 45 takes the current value of "InstrumentType" [1-128] and shifts it left by 16 bits by multiplying it with the binary number shown in Fig. 3.75. This temporary result is next saved into Parameter "InstrumentValueTemp1".

Similarly, Line 46 takes the current value of "Octave" [3] and shifts it left by 8 bits by multiplying it with the binary number shown in Fig. 3.75. This temporary result is saved into Parameter "InstrumentValueTemp2".

Line 47 adds the current values of "InstrumentValueTemp1" and "InstrumentValueTemp2" and saves this result back into "InstrumentValueTemp1" which now has the combined information on Instrument Type and Octave Level in its current value.

Line 48 adds the current value of "InstrumentValueTemp1" to "DoNote" and saves this result as "InstrumentValue1" which now has all the combined information on InstrumentType, Octave Level and Do Note. This parameter "InstrumentValue1" is used later to make the Mobile Device play a Do note at the Octave level specified and using a particular Instrument Type that would be different for each iteration of the EndLess Loop.

Line 49 adds the current value of "InstrumentValueTemp1" to "ReNote" and saved this result as "InstrumentValue2" which now has the combined information on InstrumentType, Octave Level and Re Note. This parameter "InstrumentValue2" is used later to make the Mobile Device play a Re note at the Octave level specified and using a given but different Instrument Type for each iteration of the EndLess Loop.

Line 50 does a similar work for Parameter "InstrumentValue3" which now has the combined information on InstrumentType, Octave Level and Mi Note. This parameter "InstrumentValue3" is used later by the Mobile Device to play a Mi note at the Octave level specified and using a different Instrument Type at each iteration of the EndLess Loop.

Line 51 increments "InstrumentType" by "1" to get ready for the next iteration loop. Lines 52 through 55 check on the current value of "InstrumentType" and if it is equal to "129", it is reset to "1".

Lines 56 through 61 makes the SMART Musical Instrument play each Do-Re-Mi note successively, using the current Instrument Type and Octave Level.

Video 3.32 shows how this program performs at runtime - very confusing to the ears!

3.11 More Applications with Mobile Camera: Face and Intruder Detections

The R+m.PLAY700 App has a tool for finding faces, well to be exact, it finds two blobs that can be construed as a pair of eyes! The program "FaceDetection.tskx" is a simple application of this tool (see Fig. 3.76):

The user starts the application by whistling into the mobile device (line 5).

Line 8 set the Mobile Device into a Portrait Mode.

Line 9 chooses the Back Camera.

Line 10 sets the Camera Sensor into the Face Detection Mode, and waits for the hardware to settle down by using the SMART TIMER to wait for 2 seconds (lines 11-12). Please note that the SMART TIMER resides on the Mobile Device, this timer is not the same as the CM-50's Hi-Res Timer that we had been using so far.

The Endless Loop (lines 13-23) contains the main Face Detection procedure:

Line 15 invokes the actual "Face Detection Area" software tool to run on the current image captured via the Back Camera. If among the searched 25 zones (see Fig. 3.34), it can find 2 blobs that are positioned in such a way that they can be classified as a pair of eyes, it returns that zone number which is then saved into Parameter DetectedZone. If DetectedZone has a zero value, this means that "no eyes" had been found.

```
5   WAIT WHILE ( SMART: Noise (dB) < 50 )
6
7   // Setting Back Camera to Motion Detection Mode & wait for 2 seconds
8   SMART: Screen Rotation = Portrait Mode (1)
9   SMART: Camera Selection = Back Camera (1)
10  SMART: Camera Sensor = Face Detection Mode (1)
11  SMART: Smart Timer = 2.0sec
12  WAIT WHILE ( SMART: Smart Timer > 0.000sec )
13  ENDLESS LOOP
14  {
15      DetectedZone = SMART: Face Detection Area
16      SMART: Foreground Image = 0
17      IF ( DetectedZone != 0 )
18      {
19          SMART: Foreground Image = [Position:DetectedZone],[Item:1],[Size:0],[Color:Unknown]
20          High-resolution Timer = 0.125sec
21          WAIT WHILE ( High-resolution Timer > 0.000sec )
22      }
23  }
```

Fig. 3.76 Face Detection Application.

Line 16 clears the display of previous foreground images.

Line 17 uses an IF structure to test if DetectedZone is non-zero, i.e. if "a pair of eyes" had been found. If this is TRUE, then line 18 would display the Foreground Image Item 1 (see Fig. 3.77) in the DetectedZone's location using the original size and colors. If FALSE, program flow control is transferred back to the beginning of the EndLess Loop (line 14).

Lines 20-21 use the CM-50 Hi-Res Timer to delay for 0.125 second (so that the Foreground Image Item 1 stays on for a little while). A good question is whether the SMART TIMER can be used here instead? See Video 3.33 for the runtime performance and more discussions.

Fig. 3.77 Foreground Image Item 1 used in Face Detection Application.

Another R+m.PLAY700 tool for the video camera is the Motion Detection tool that can be applied to an Intruder Alert project. The solution program is "IntruderAlert.tskx" (see Fig. 3.78):

The user starts the application by whistling into the mobile device (line 5).

Line 8 set the Mobile Device into a Portrait Mode.

Line 9 chooses the Back Camera.

Line 10 sets the Camera Sensor into the Motion Detection Mode, and waits for the hardware to settle down by using the SMART TIMER to wait for 2 seconds (lines 11-12). Please again note that the SMART TIMER resides on the Mobile Device, this timer is not the same as the CM-50's Hi-Res Timer that we had been using so far.

```
5   WAIT WHILE ( SMART: Noise (dB) < 50 )
6
7   // Setting Back Camera to Motion Detection Mode & wait for 2 seconds
8   SMART: Screen Rotation = Portrait Mode (1)
9   SMART: Camera Selection = Back Camera (1)
10  SMART: Camera Sensor = Motion Detection Mode (3)
11  SMART: Smart Timer = 2.0sec
12  WAIT WHILE ( SMART: Smart Timer > 0.000sec )
13
14  ENDLESS LOOP
15  {
16    DetectedZone = SMART: Motion Detection Area
17    IF ( DetectedZone != FALSE (0) )
18    {
19      SMART: Photo Capture = Capture (1)
20      Buzzer Timer = Melody Time
21      Buzzer Index = Melody No.20 (20)
22      WAIT WHILE ( Buzzer Timer > 0.000sec )
23      WAIT WHILE ( SMART: Photo Capture != 0 )
24    }
25  }
```

Fig. 3.78 TASK Code for Intruder Alert Application.

The Endless Loop (lines 15-25) contains the main Motion Detection procedure:

Line 16 invokes the actual "Motion Detection Area" software tool to run on the current and previous images captured via the Back Camera. If among the searched 25 zones (see Fig. 3.34), it can find a zone whereas pixels have changed enough in their color values between the current and previous images, it would return that zone number which is then saved into Parameter DetectedZone. If DetectedZone has a zero value, this means that "no motion" had been detected.

Line 17 uses an IF structure to test if DetectedZone is non-zero, i.e. if "motion" had been detected. If this is TRUE, then line 19 would capture the current image with the intruder hopefully still in it, and the CM-50 plays Melody 20.

Line 23 waits until the process of image capturing and image saving onto the mobile device is really done before the CM-50 can perform another iteration of the EndLess Loop.

Video 3.34 shows the runtime performance of this application.

Recapping this Chapter, TASK is a "compiled" language meaning that although the user edits the robot program on the PC or Mobile Device in a human-friendly way, the TASK tool next converts it into a "machine language" version that runs on the CM-50 Controller. Thanks to this runtime environment, the access to sensors/actuators and the robot control algorithm are at their best performance as all the hardware and software involved are "physically" near to each other (see Fig. 1.7). For Multimedia needs, TASK relies on the R+m.PLAY700 App which brings in the Mobile Device as a powerful co-controller. The linkage between the CM-50 and the Mobile Device relies on Bluetooth communications at a rate of 57 Kbps which cannot be modified by the user.

Currently the support materials for the R+m.PLAY700 are only in Korean: (http://support.robotis.com/ko/software/mobile_app/r+smart/smanrt_manual.htm, http://support.robotis.com/ko/software/roboplus2/r+task2/programming3_paramerter/r+task2_parameter2.htm and http://support.robotis.com/ko/software/roboplus/roboplus_task/programming/parameter/remoteterminal/smart_parameter.htm), however if the reader uses Chrome for web browser, these pages can be translated into fairly "usable" English. The author also includes with this book a PDF file named "SmartCommandsControlTable.pdf" which essentially is a Chrome-translated version of the on-line Table. There are many more interesting and useful tools available in the R+m.PLAY700 App, the reader is encouraged to try them all.

The next Chapter deals with SCRATCH which is an "interpreted" language meaning that the user edits the program and runs it at the PC level, and all ROBOTIS hardware commands are routed to the CM-50 via a helper application called R+SCRATCH (see Fig. 1.8). This adds another hardware/software layer between the robot control algorithm (which runs on the PC) and the actual robot, which affects the runtime performance of the robot sometimes adversely. On the other hand, as the PC's Operating System is much more powerful and sophisticated than the one on the CM-50, multiple parallel tasks can be enabled inside a SCRATCH program while this feature is not supported by TASK. The linkage between the CM-50 and the PC/R+SCRATCH also relies on Bluetooth communications set at a fixed rate of 57 Kbps.

3.12 References

Bergin J (2012) Pedagogical patterns: advice for educators. CreateSpace Independent Publishing Platform, p. 115.

Thai CN (2017) Exploring Robotics with ROBOTIS Systems (2nd Edition). Springer Publishers International, New York.

3.13 Review Questions for Chapter 3

1. How often is the CALLBACK Function executed during runtime of a TASK program?
2. List the bit-level operators available in the TASK tool.
3. Does the CM-50 support floating-point arithmetics?
4. List 3 types of LOOPS in TASK.
5. From which side of a LOAD statement, does it begin to execute?
6. Is there any difference between the operators "=" and "=="?
7. How many Musical Instruments are allowed in the R+m.PLAY700 App?
8. How many Audio Sources are available to use in the R+m.PLAY700 App?
9. Describe the SMART CONSTANT structure used for Musical Instruments in the R+m.PLAY700 App.
10. Describe 3 programming techniques that can be used for programming robot autonomous behaviors.
11. Describe 3 ways to achieve Remote Control of robots.

Chapter 4: Using R+SCRATCH & SCRATCH® 2

MIT's SCRATCH 2 software is a very popular software platform with millions of users worldwide (https://scratch.mit.edu/), and reference texts on this subject are too numerous to list in this chapter. The author would recommend such works as Ford (2014) and Warner (2015) for self-learners who want to have a thorough understanding of the SCRATCH 2 language. If the reader is a teacher/instructor then the author would also recommend the book by Bagge (2015) and his Code-IT web site (http://code-it.co.uk/philbagge). In this Chapter, only selected SCRATCH 2 features, most applicable to robotics, would be presented.

ROBOTIS designs the software tool R+SCRATCH to be a helper application that links the SCRATCH 2 IDE directly to the CM-50 Controller's Firmware (see Figs. 1.8 and 4.1). This means that any TASK code previously downloaded onto the CM-50 is disabled (not erased) as soon as R+SCRATCH connects to the CM-50 (see Part 1 of Video 4.1).

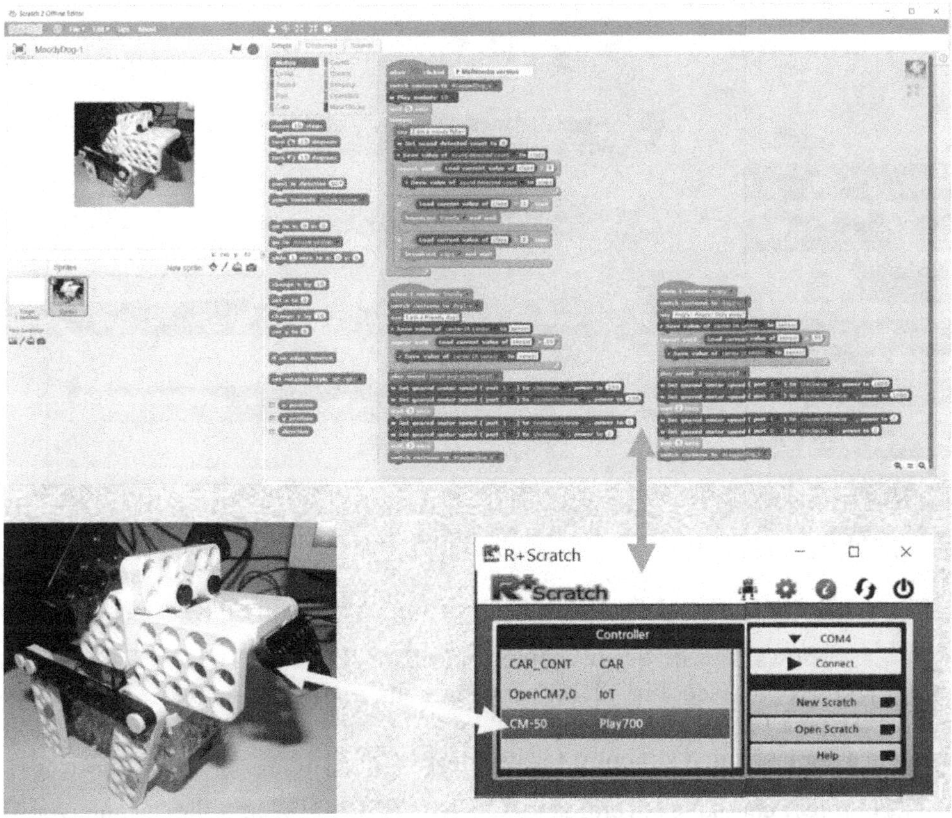

Fig. 4.1 Information Flow between SCRATCH 2, R+SCRATCH and CM-50 Controller.

ROBOTIS also creates a custom SCRATCH HTTP Extension tool (https://wiki.scratch.mit.edu/wiki/Scratch_Extension) to the SCRATCH 2 Off-Line Editor to make available 8 SCRATCH Block Codes that can interface with the CM-50's firmware (see Fig. 4.2). The three "Play" and the three "Set" Block Codes are of the "w" type, i.e. a "write" type of command. Each "write" command requires 50 ms delay to progress, from the SCRATCH 2 IDE through the R+SCRATCH helper application, and all the way down to the CM-50 Firmware and actuators mounted on the PLAY700 robot. There is only one "r" (i.e. "read") type of command, "Save value of", which has a pull-down menu to access two types of sound counter and the three built-in IR Sensors (Left, Center, Right). Each "read" command requires 70-100 ms delay to process, once it is triggered at the SCRATCH 2 IDE level. When using TASK codes as shown in Chapter 3, these types of commands cost much less time to be performed (less than 5 ms), so the much longer time delays when using SCRATCH 2 create important and sometimes adverse side effects (programming and physical) affecting a robot runtime performance. These effects will be shown and discussed further in later parts of this Chapter.

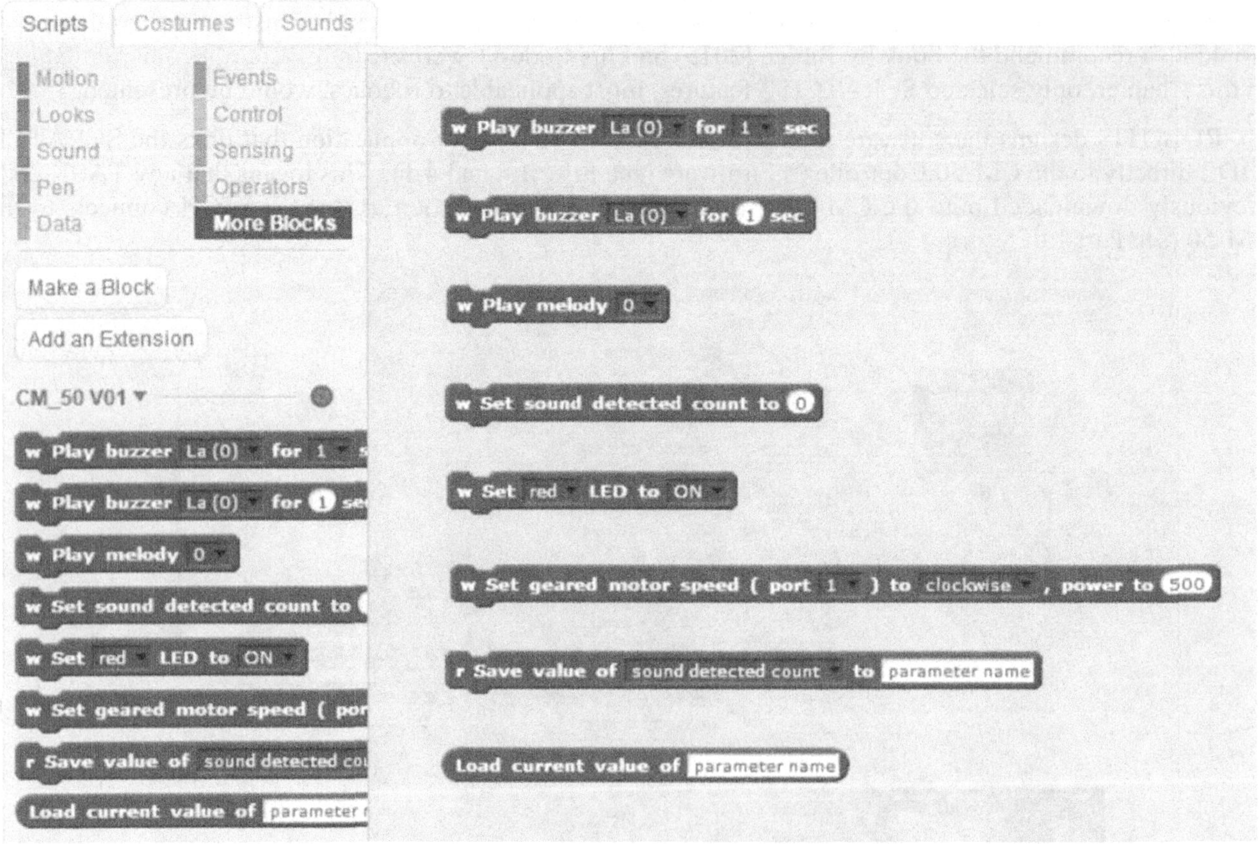

Fig. 4.2 ROBOTIS CM-50 Block Codes compatible with SCRATCH 2 IDE.

Part 2 of Video 4.1 shows a basic usage of the tool chain R+SCRATCH/SCRATCH 2 Offline Editor and the "Spinning Top" robot (see Fig. 2.10), resulting in a project named "Maneuvers-1.sb2" to be described in Section 4.1 (see Fig. 4.3). Although SCRATCH 2 is designed for beginner programmers aged 8 or older, it has quite a sophisticated structure for its "project":

Each SCRATCH project can have several actors called "SPRITES" and the "Maneuvers-1" project uses a single Sprite called "Sprite 1" (see Fig. 4.3 – bottom left panel). The large area on the top left panel, where a larger picture of Sprite 1 can be found, is called the STAGE.

Attached to each Sprite are SCRIPTS (actual codes), its COSTUMES (graphics components triggered by its SCRIPTS), and its SOUNDS (audio components also controlled by its SCRIPTS) – see top of mid-panels of Fig. 4.3.

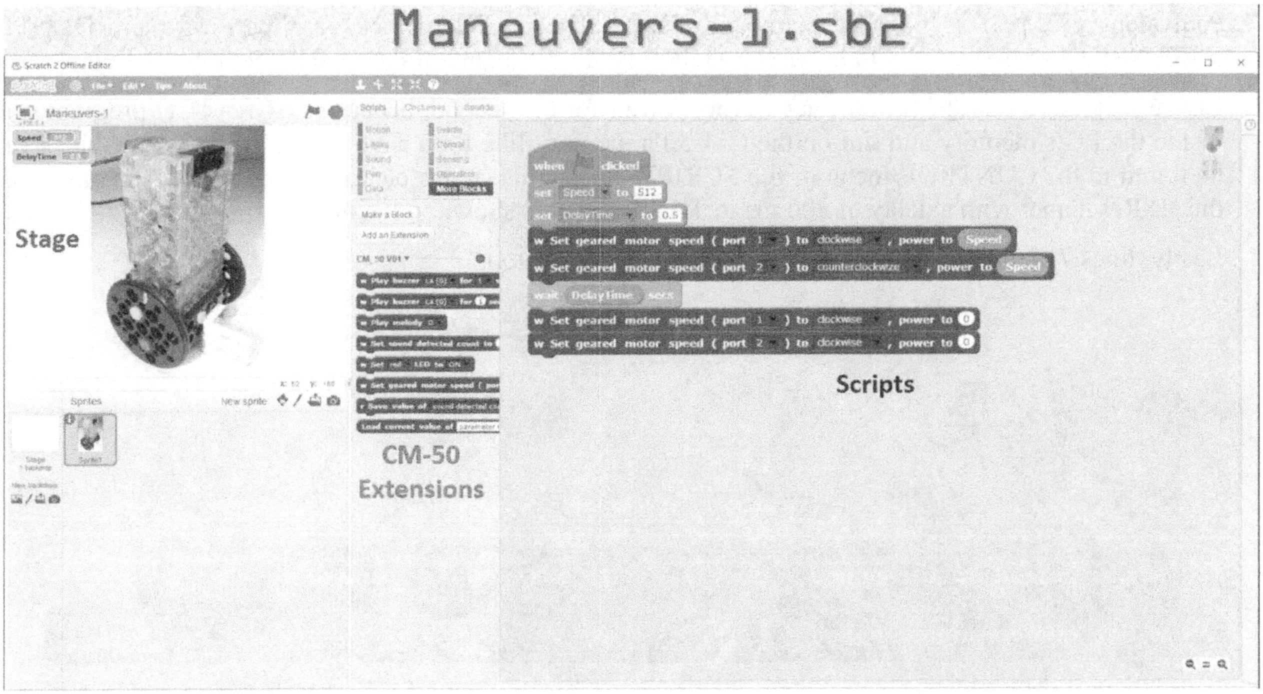

Fig. 4.3 Components of the "Maneuvers-1.sb2" SCRATCH 2 project.

To allow SCRATCH 2 to interact with the robot via R+SCRATCH, the user must also mount appropriate "EXTENSION" packages (i.e. a *.JSON file – see bottom of mid-panel in Fig. 4.3). Please see Video 4.2 for a primer on how to create and edit a SCRATCH 2 project that would interact with this example robot.

The right panel contained the actual project's SCRIPTS which were created using different types of colored blocks (representing different data and behavior structures) that the user could stack and unstack to build the logic required for a project.

4.1 Basic Robot Programming with SCRATCH

In this section, the goal is to show the similarities and differences in creating and running 2 pieces of codes on the same robot with the same required physical behavior (i.e. "go forward for 0.5 second"). One is a TASK code from Section 3.2 named "Maneuvers-1.tskx" and the other is a SCRATCH program called "Maneuvers-1.sb2" (see Video 4.3 for complete steps).

Fig. 4.4 displays the MAIN SCRIPT which starts when the user clicks on the GREEN FLAG block (which has similar functions as the START PROGRAM statement used in TASK codes):

The GREEN FLAG block can be found in the EVENTS menu located in the SCRIPTS tab.

Next, Variables "Speed" and "DelayTime" are set respectively to "512" and "0.5" (lines 1-2, i.e. same functionality as Assignment statements inside TASK). The Variable parameters can be created and set using items found in the DATA menu of the SCRIPTS tab.

Then, Port 1 Motor is set to a Clockwise motion with a power equaled to Variable Speed (line 4), while Port 2 Motor is set to a Counter-Clockwise motion with a power also equaled to Speed (line 5). Together, these two Code Blocks make the Bot go forward. Inside TASK codes, we would use an equivalent COMPUTE type of command here. These Code Blocks are part of the ROBOTIS EXTENSION so they can be found in the MORE BLOCKS menu of the SCRIPTS tab.

Line 6 just makes the PC (not the actual robot) wait for 0.5 second, because a SCRATCH program runs inside the PC's memory and **not** on the CM-50's memory like for a TASK code. The WAIT block can be found in the CONTROL menu of the SCRIPTS tab. This step is clearly easier to use than setting up the Hi-Res Timer with a delay of 500 ms in TASK codes as shown in Section 3.2.

Lastly, lines 7 and 8 turn off the motors to make the robot stop.

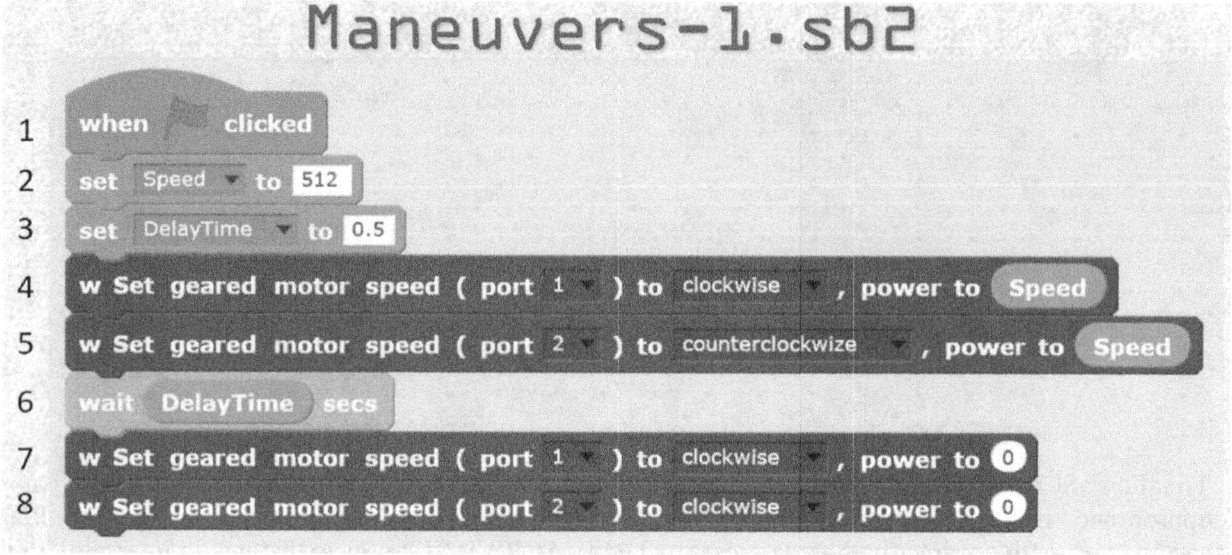

Fig. 4.4 MAIN SCRIPT for "Maneuvers-1.sb2" project.

Video 4.3 also contrasts the runtime performances of this robot when running a SCRATCH code (Maneuvers-1.sb2) and when running an equivalent TASK code (Maneuvers-1.tskx). For the SCRATCH version, the unexpected "jerks" at the beginning and end of the robot's forward motion are due to the longer delays for a SCRATCH command to go from the PC to the robot Controller to activate a given motor. Essentially, Port 1 Motor (on robot's right) was activated/deactivated about 50 ms before Port 2 Motor (on robot's left) and thus created the beginning "left" jerk and the final "right" jerk. Logic-wise, the same timing delays are present in the TASK code, but because it runs directly inside the CM-50 controller's memory space, and therefore these commands can reach the right and left motors much faster. This explains why we did not witness these effects when using TASK.

4.2 Sequential Control of Robot Maneuvers

When using TASK, back in Section 3.2, we learned how to create independent program modules called FUNCTIONs which were then had to be CALLed from inside the MAIN Task program to execute these functions (see Figs. 3.5 and 3.6). In SCRATCH, ones can create similar independent code-block modules as shown in "Maneuvers-3.sb2" (see Fig. 4.5). These SUB-SCRIPTs were duplicated from the single one found in "Maneuvers-1.sb2" (Fig. 4.4) and individually modified with the appropriate turn directions, clockwise or counter-clockwise, to create four robot maneuvers: Going Forward, Going Backward, Turn Left and Turn Right. To activate any SUB-SCRIPT, the user only needs to click on the appropriate GREEN FLAG block and the robot performs the user-wanted maneuver (see Video 4.4).

The author is quite certain that the reader already notices that this is a completely new and **interactive** way to think about problem-solving, code generation, execution and trouble-shooting when using SCRATCH, as compared to the longer cycle of "problem-solving/editing/compiling/downloading/executing/trouble-shooting" needed when using TASK. This is the first SCRATCH feature that promotes a "parallel/lateral thinking" mode for a SCRATCH practitioner.

However, in most robotics applications, autonomous behavior is needed, thus let's first see how "sequential control" can be implemented in SCRATCH. The resulting program is "Maneuvers-4a.sb2" and it is shown partially in Fig. 4.6 for the MAIN SCRIPT and the two SUB-SCRIPTs "Forward" and "Stop". The matching TASK program is called "Maneuvers-4.tskx" (see Figs. 3.7 and 3.8). The goal is to make the robot perform 3 maneuvers in the following order: Go Forward, Turn Right, Go Backward and then stops.

The project "Maneuvers-4a.sb2" (Fig. 4.6) implements another powerful "parallel-thinking" feature, available only in SCRATCH, which is the ability to BROADCAST and RECEIVE messages among various SCRIPTS. These types of code-blocks can be found in the menu EVENTS in the SCRIPTS tab of the SCRATCH IDE.

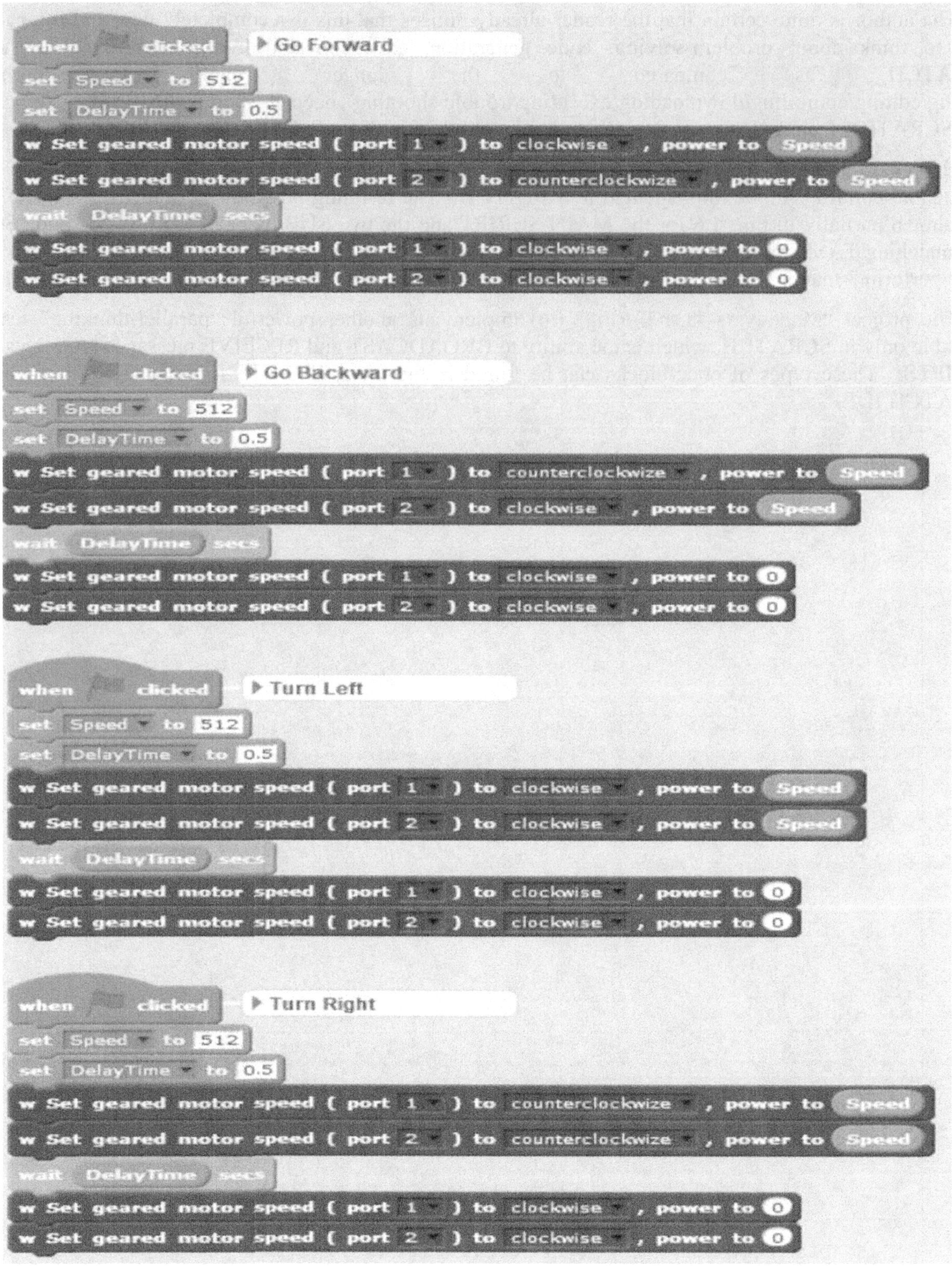

Fig. 4.5 Independent Maneuvers SUB-SCRIPTs for "Maneuvers-3.sb2" project.

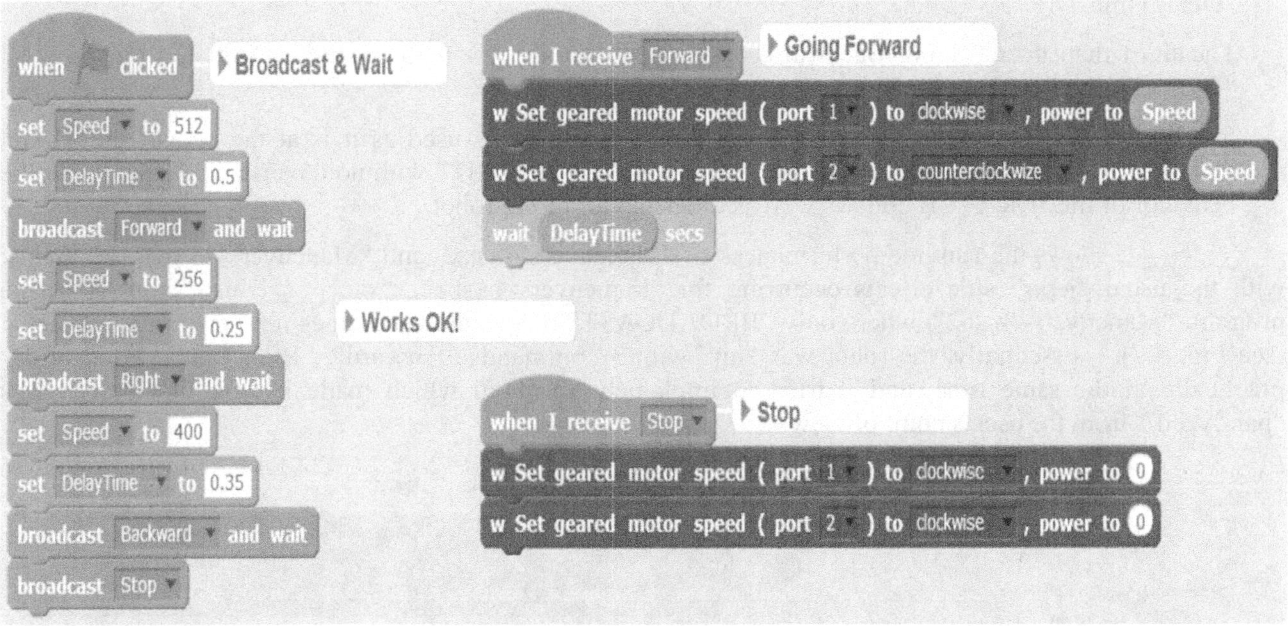

Fig. 4.6 MAIN SCRIPT and two Maneuvers SUB-SCRIPTs in "Maneuvers-4a.sb2" project.

Let's have a closer look at the SUB-SCRIPT "Forward" (top right in Fig. 4.6). The "When I Receive" type of block can be found in the menu "EVENTS" of the SCRIPTS tab and it should be the FIRST block for this type of SUB-SCRIPT. It is the equivalent of the FUNCTION DEFINITION statement in a TASK code which likewise marks the beginning of a Function (see Section 3.2). Next are the two commands to turn on the motors on Port 1 and Port 2, then the "WAIT for DelayTime" block. However, the previous two commands to turn off the motors (see Fig. 4.5) are now set up in their own SUB-SCRIPT called "Stop" - as another example of code modularization (bottom right in Fig. 4.6).

The MAIN SCRIPT starts from the top left in Fig. 4.6 with the GREEN FLAG block:

First, Speed and DelayTime were set to "512" and "0.5" respectively.

Then the "BROADCAST and WAIT" block is used to send out the message "Forward" which is of course picked up by the SUB-SCRIPT "Forward". Here there are 2 parallel and simultaneous actions that happen: a) the SUB-SCRIPT "Forward" executes its blocks which makes the robot go forward at speed 512 and for 0.5 second; b) the MAIN SCRIPT is still "live" but as its current block is a "BROADCAST & WAIT" type, it now enters its WAIT phase. When the SUB-SCRIPT "Forward" is done, this SUB-SCRIPT terminates itself, then the MAIN SCRIPT can go on to the next block, which is to set the Variable Speed to a new value of "256". These steps are similar to the ones taken by a TASK program when it is calling a Function (section 3.2), however there is only 1 program flow control pointer used in a TASK code, while SCRATCH can keep track of "multiple" program flow control pointers (i.e. multiple BROADCAST messages, all thanks to the PC powerful multi-tasking Operating System).

After setting Speed and DelayTime to their new values of "256" and "0.25", the MAIN SCRIPT issues another "BROADCAST & WAIT" block, this time to the SUB-SCRIPT "Right" which makes the robot turns right at speed 256 and for 0.25 second.

Next, the MAIN SCRIPT sets the new values of "400" and "0.35" respectively to "Speed" and "DelayTime".

The third maneuver "Going Backwards", at speed 400 and for 0.35 second, is initiated by the third "BROADCAST & WAIT" block.

Finally, to stop the bot, a plain "BROADCAST" command is used as it is at the end of the MAIN SCRIPT. Here, the reader can also use "BROADCAST & WAIT" with no adverse consequence to the behavior of the SCRATCH code as well as the behavior of the robot.

Video 4.5 shows the runtime performances of "Maneuvers-4.tskx" and "Maneuvers-4a.sb2" programs, with the usual "jerks" side effects occurring for "Maneuvers-4a.sb2". Video 4.5 also shows that the program "Maneuvers-4b.sb2", where only "BROADCAST" blocks are used, does not perform as expected (see Fig. 4.7). Essentially, the robot was "hit" with 3 commands "Forward", "Right" and "Backwards" practically at the same time, and it tried to implement them all which made it look like that it was "paralyzed", from the user's point of view.

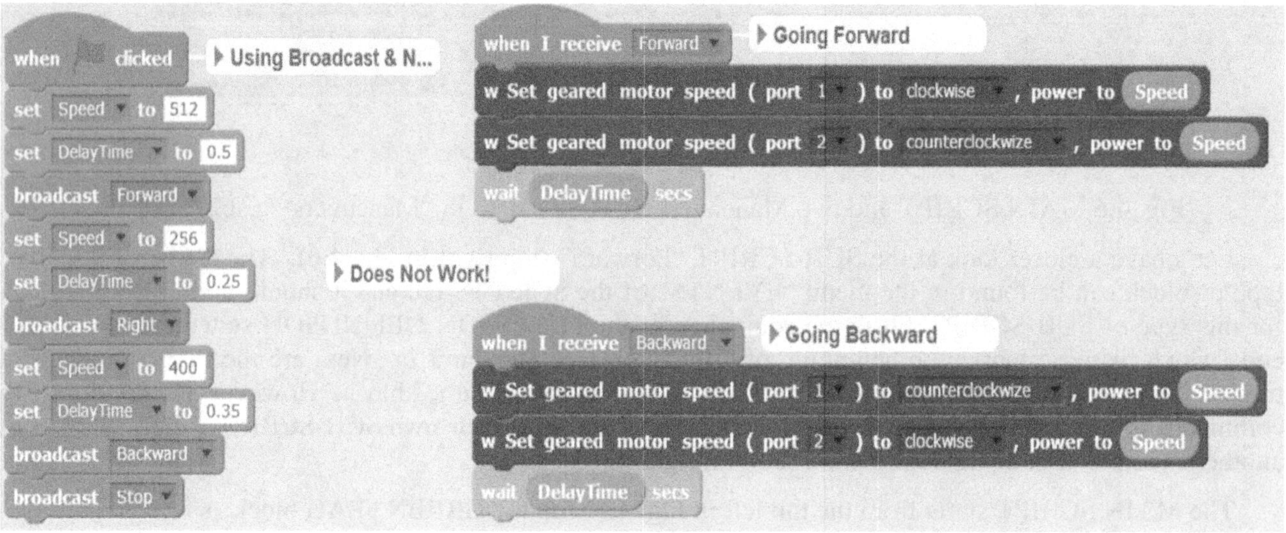

Fig. 4.7 MAIN SCRIPT and two Maneuvers SUB-SCRIPTs in "Maneuvers-4b.sb2" project.

4.3 "Spinning Top" Maneuvers using on Center IR Sensor

In this section, we take another step towards robot autonomous control with SCRATCH. The program "IR-Sensor-Speed-1.sb2" links the real-time value of the Center IR Sensor to the Forward motion of the robot (see Fig. 4.8). Not shown in this Fig. 4.8 are the definitions of the Variables "Speed" and "IR" via the DATA menu in the SCRIPTS tab.

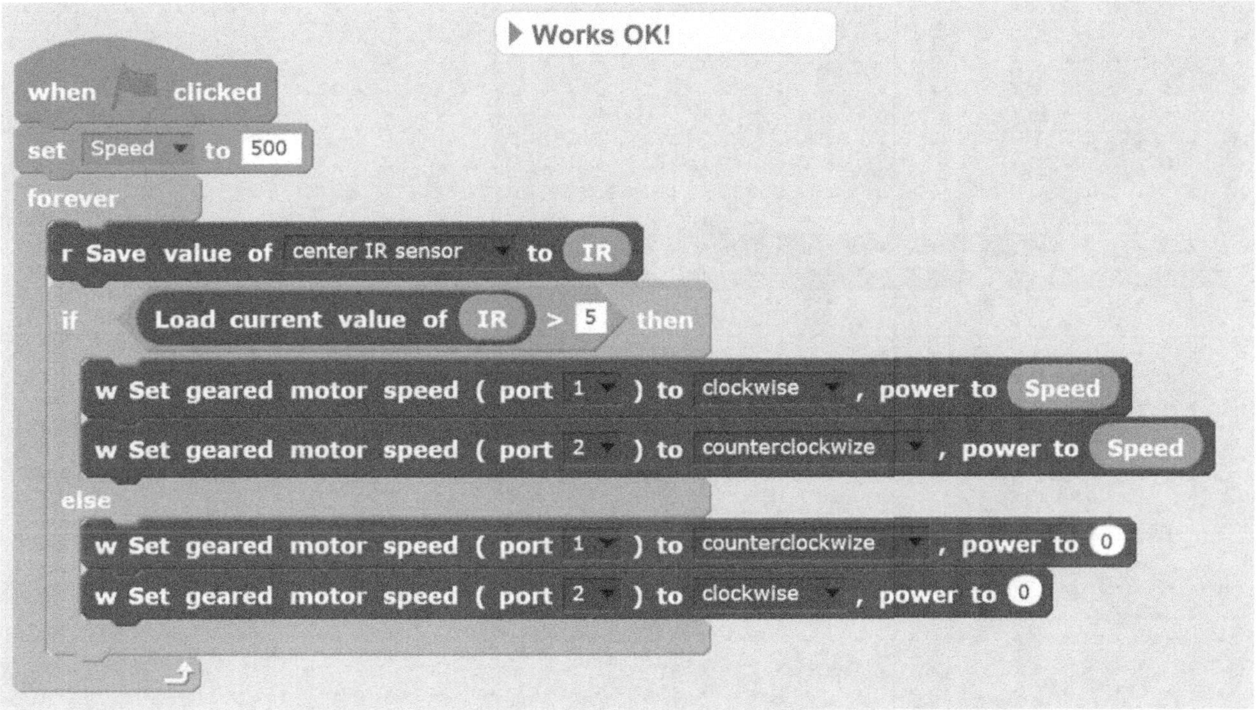

Fig. 4.8 MAIN SCRIPT for "IR-Sensor-Speed-1.sb2" project.

It uses a FOREVER LOOP (same behavior as an ENDLESS LOOP in TASK) which can be found in the menu CONTROL of the SCRIPTS tab:

For each iteration of this FOREVER LOOP, a "SAVE VALUE OF" block is issued to the CM-50 Controller to return the current reading of its Center IR Sensor and save it in the Variable IR.

Next, an IF-ELSE structure is used (it can also be found in the CONTROL menu). The conditional expression of the IF part is created using the ">" block found in the OPERATORS menu combined with the "LOAD CURRENT VALUE OF" block which is accessible from the "MORE BLOCKS" menu. The Variable IR block is also needed and it is accessible from the DATA menu. Video 4.6 has the detailed steps for creating such an IF-ELSE structure.

During any iteration when this conditional expression evaluates to TRUE, the robot is made to go forward at Speed 500. Otherwise, the robot stays still.

The runtime behavior of this program/robot is also shown in Video 4.6: when the robot detects an object close enough its "head", it would dart forward to get away, then it would stop - ready to play again!

Next, the author would like to point out a possible syntax error that the users do not want to make, as it took the author quite a bit of time to figure this problem out. Fig 4.9 shows the SCRATCH program "IR-Sensor-Speed-2.sb2" which does **not work** at all, even though no syntax error was detected. Essentially the Variable "IR" is used directly in the IF statement, as this would be a sensible thing to do, considering that the code already had "Saved value of center IR sensor to IR" in the preceding block. So, the "take-home" lesson is to use the block "Load current value of" whenever ones need to access the value of a Variable that was obtained via a "Save value of" block.

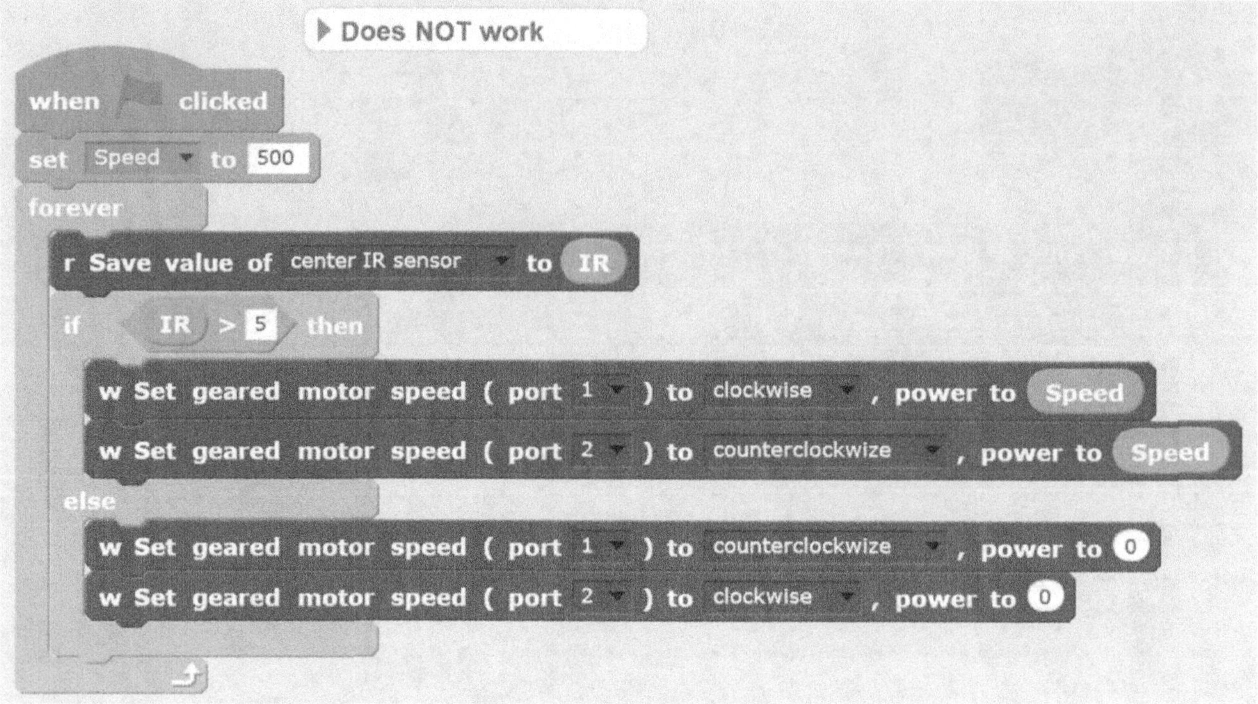

Fig. 4.9 MAIN SCRIPT for "IR-Sensor-Speed-2.sb2" project.

Now we are ready for our first implementation of the Sense-Think-Act paradigm previously discussed in Chapter 1 to the SCRATCH environment. The sensing component will be the IR Center Sensor of the "Spinning Top" robot as before.

The tasks assigned to this robot are as follows (review Section 3.3 also):

At start, the robot plays a musical melody (Melody 2) to let the user know that it is "ready for action", and its Speed initialized to zero.

Next, the behaviors of the robot depend on the position of an object relative to the IR Center Sensor, thus the condition-action list to be considered is as follows:

If the IR Center Sensor reading is less than 50, the robot stays still.

If the IR Center Sensor reading is between 50 and 300, the robot spins right with its speed proportional to the IR Center Sensor reading and it plays a musical instrument for 0.5 beat.

If the IR Center Sensor reading is between 300 and 500, the robot spins left with its speed proportional to the IR Center Sensor reading and it plays a second and different musical instrument for 0.5 beat.

If the IR Center Sensor reading is larger than 500, the robot rolls forward at Speed=512 to escape from the object and it plays a third musical instrument for 0.5 beat.

This condition-action list indicates that we are dealing with a situation whereas one and only one condition can happen at any one time during the runtime performance of this robot, thus we will need to use an IF-ELSE-IF structure for our solution to this problem.

The program "IR-based-Maneuvers.sb2" describes the solution approach used:

The Initial Event is solved by playing Melody 2 out of the CM-50 speaker and Variable Speed is initialized to zero (see Fig. 4.10).

Fig. 4.10 Initial Event for program "IR-based-Maneuvers.sb2".

The main algorithm (Fig. 4.11) uses an outer FOREVER loop to represent the World-Robot Interactions Cycle (previously depicted in Fig. 1.10). This loop contains an IF-ELSE-IF structure representing the condition-action/input-output listing as detailed out previously.

During each loop, the PC would evaluate/execute this IF-ELSE-IF structure in the following manner:

It acquires the latest IR Center Sensor reading and saves it in Variable IR_Sensor.

Next, it enters the IF-ELSE-IF structure and checks the first IF statement, i.e. the condition (IR_Sensor < 50). If this condition is TRUE, it BROADCASTs the message "Stop" and WAITs for the execution of the SUB-SCRIPT "Stop" to finish, it then exits the whole IF-ELSE-IF structure and performs another iteration of the FOREVER loop. But if this condition is FALSE, i.e. (IR_Sensor >= 50), the PC goes on to check the next second IF statement which is based on the condition (IR_Sensor < 300).

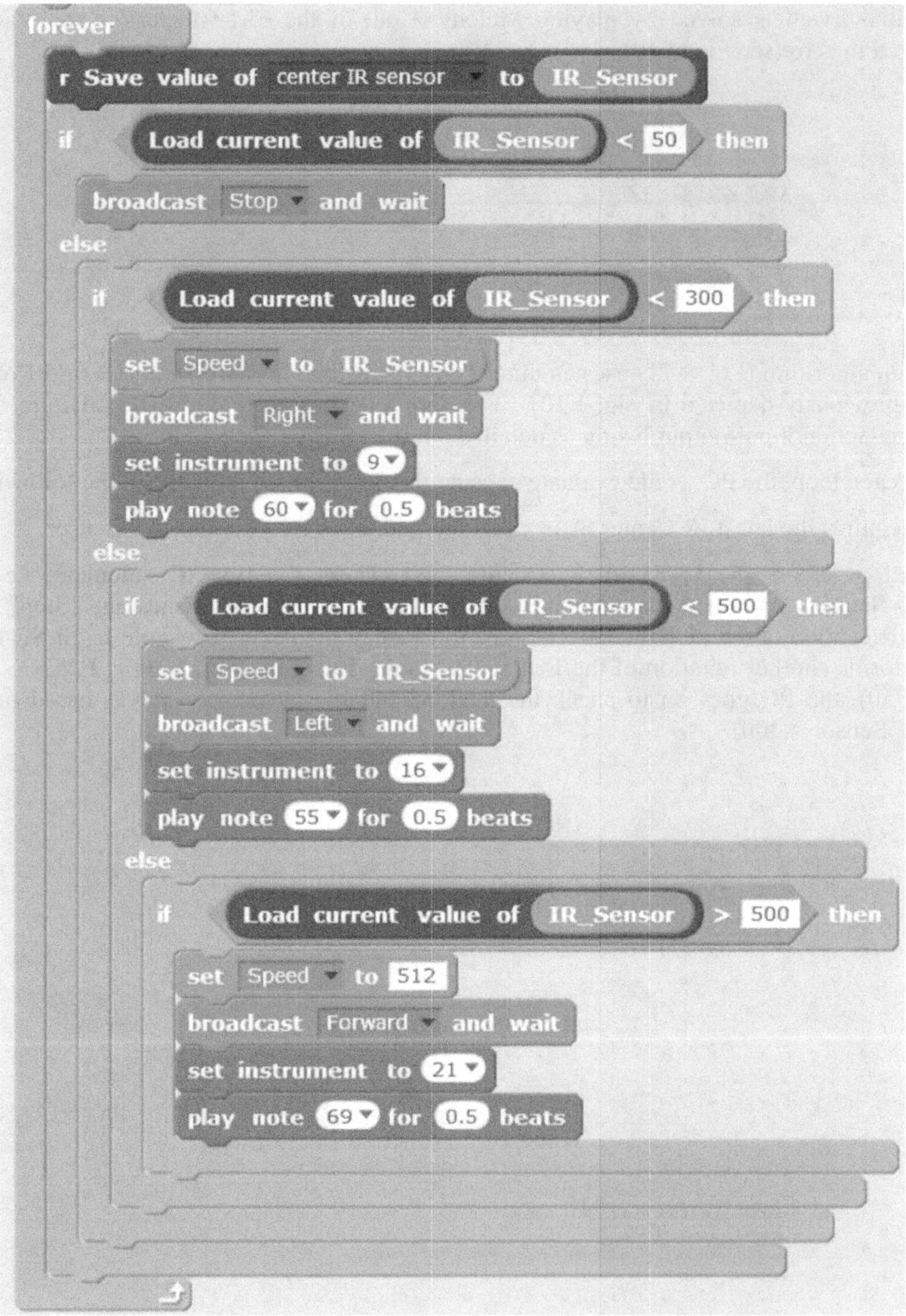

Fig. 4.11 Main Algorithm used in program "IR-based-Maneuvers.sb2".

Please note that according to the considered condition-action listing, the PC is supposed to check if IR_Sensor is between 50 and 300, but here it uses a condition that involves only checking for (IR_Sensor < 300). Is that sufficient to do what is required? And the answer is Yes and that it is

OK to just check for the condition (IR_Sensor < 300) here, because the condition (IR_Sensor >= 50) is implicitly TRUE when the condition (IR_Sensor > 50) evaluates to FALSE (see first IF).

Thus, if the second condition (IR_Sensor < 300) turns out to be TRUE (meaning that IR_Sensor is between 50 and 300), the PC assigns the value of IR_Sensor to Speed and BROADCASTs & WAITs for SUB-SCRIPT "Right" to make the robot spin right at a Speed proportional to IR_Sensor's value. The PC also plays Note 60 on Instrument 9 for 0.5 beats, then it exits this IF-ELSE-IF structure and makes another iteration of the FOREVER loop. However, if the condition (IR_Sensor < 300) turns out to be FALSE, i.e. (IR_Sensor >= 300), then the PC goes on to check the third IF statement.

Similarly, with the third IF, the PC is checking to see if IR_Sensor is between 300 and 500. If this is TRUE, then the robot spins left with a Speed proportional to IR_Sensor's value, plays Note 55 on Instrument 16 for 0.5 beats, exits this IF-ELSE-IF structure and continues with a new iteration of the FOREVER loop. If the condition of the third IF turns out to be FALSE (i.e. (IR_Sensor >= 500), then the fourth IF is executed next.

If the fourth IF, i.e. (IR_Sensor > 500), turns out to be TRUE, then the PC would run forward at Speed=512 and plays Note 69 on Instrument 21 for 0.5 beats. This maneuver is designed to make the robot run away from under the approaching object. Once away from the object's influence, the IR_Center Sensor would read zero, and the robot would stop and be ready to "play again".

Video 4.7 gives a verbal explanation of the program "IR-based-Maneuvers.sb2" and shows its runtime performance.

Fig. 4.12 shows the program "IR-based-Maneuvers-Say.sb2" which is based on the previous program "IR-based-Maneuvers.sb2", with appropriate "Text Balloons" generated with SAY blocks from the LOOKS menu in the SCRIPTS tab. Video 4.8 shows the runtime performance of this program.

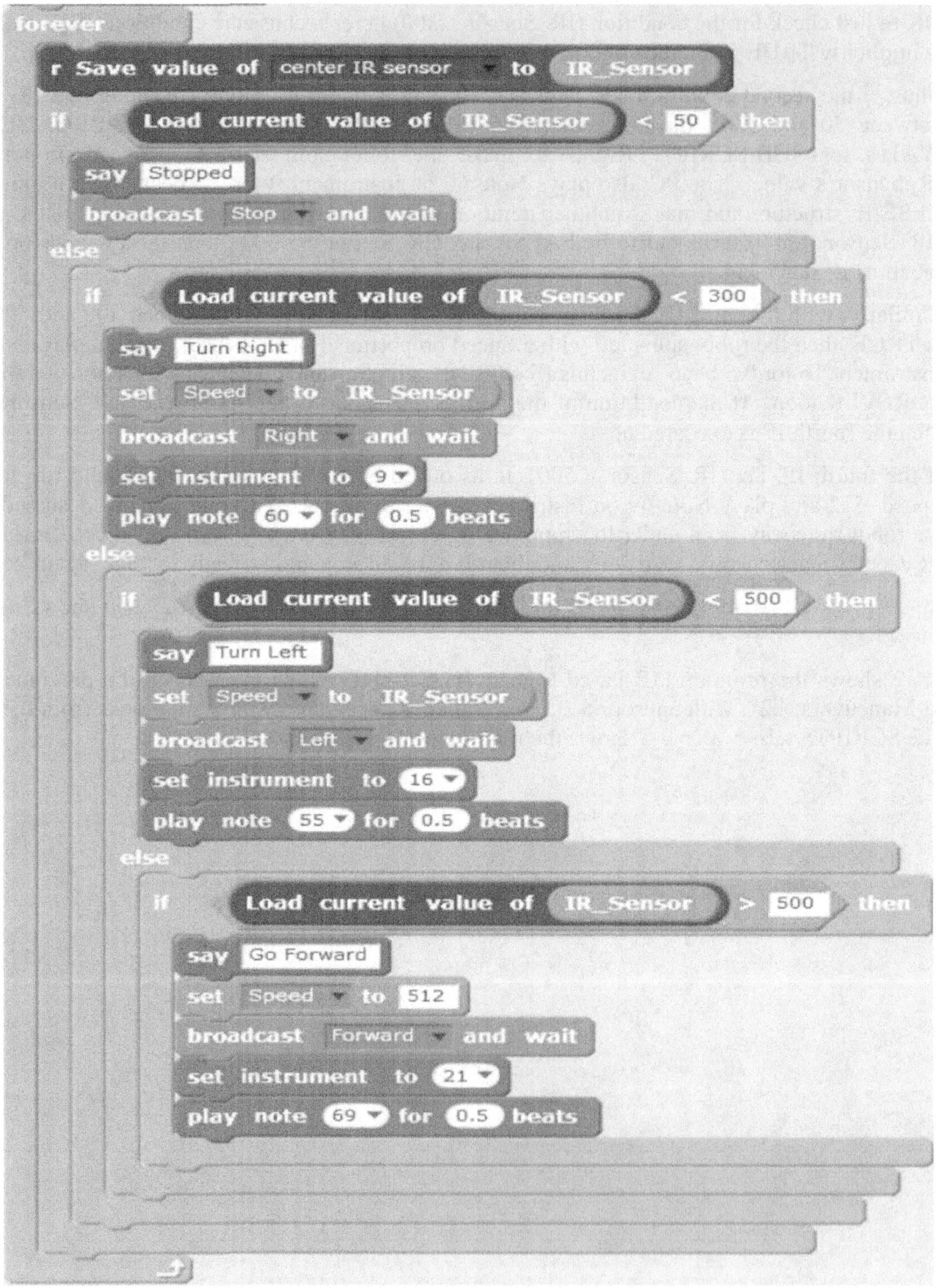

Fig. 4.12 Main Algorithm used in program "IR-based-Maneuvers-Say.sb2".

4.4 "Avoider" Maneuvers using Left and Right IR Sensors

In this section, the previous "Spinning Top" robot is still being used, but it is renamed "Avoider" to emphasize the types of maneuvers described herein. The "Avoider" will be using its Left and Right IR Sensors to detect a single object and to accomplish tasks described in the following condition-action list:

If there is no obstacle in front of the Avoider, it can go forward.

If there is an object on the Left of the Avoider, it should back away and turn to the Right to avoid the obstacle.

If there is an object on the Right of the Avoider, it should back away and turn to the Left to avoid the obstacle.

If there is an object in Front of the Avoider, e.g. a Corner Wall, it should first go Straight Back to avoid the obstacle then turn Left and Backward to steer away from the obstacle.

This is another familiar programming situation of **choosing one and only one alternative among several possible alternatives**, thus an **IF-ELSE-IF** structure should do an adequate job when combined with an **Outer FOREVER Loop**. The SCRATCH solution is named "Avoider-2.sb2". It starts with the Playing of Melody 2 and the SETting of Variable Speed to 256. The main algorithm is described in Fig. 4.13:

The FOREVER Loop starts out by reading in the latest values from the IR Left and Right Sensors and saves them respectively into Variables IR_Left and IR_Right.

Next is an IF-ELSE-IF structure with 4 mutually exclusive cases:

For the First IF block, if the values of "IR_Left" AND "IR_Right" are both less than 1 (i.e. no obstacle in front of Avoider) then a "Go Forward" Text Balloon pops up in the Stage area and the message "Forward" is BROADCASTed to the SUB-SCRIPT "Forward" which makes the robot go forward. The PC then exits the IF-ELSE-IF structure and begins another iteration of the FOREVER loop. However, if this condition evaluates to FALSE, meaning that either "IR_Left" or "IR_Right" have values larger than 1, or in other words when **either** IR sensor sees an object, the PC would need to check further by executing the Second IF block.

In the Second IF block, the condition ((IR_Left > 20) AND (IR_Right < 1)) represents a situation where there is an object on the Left of the robot and none on its Right side. If this condition evaluates to TRUE, a "Right & Back" Text Balloon appears in the Stage and the robot turns Right by Backing away to avoid the obstacle on its left (i.e. block (**BROADCAST "RightBack" and Wait**)). The PC next exits the IF-ELSE-IF structure, and starts the next iteration of the FOREVER loop. If this condition evaluates to FALSE, then the Third IF block is executed.

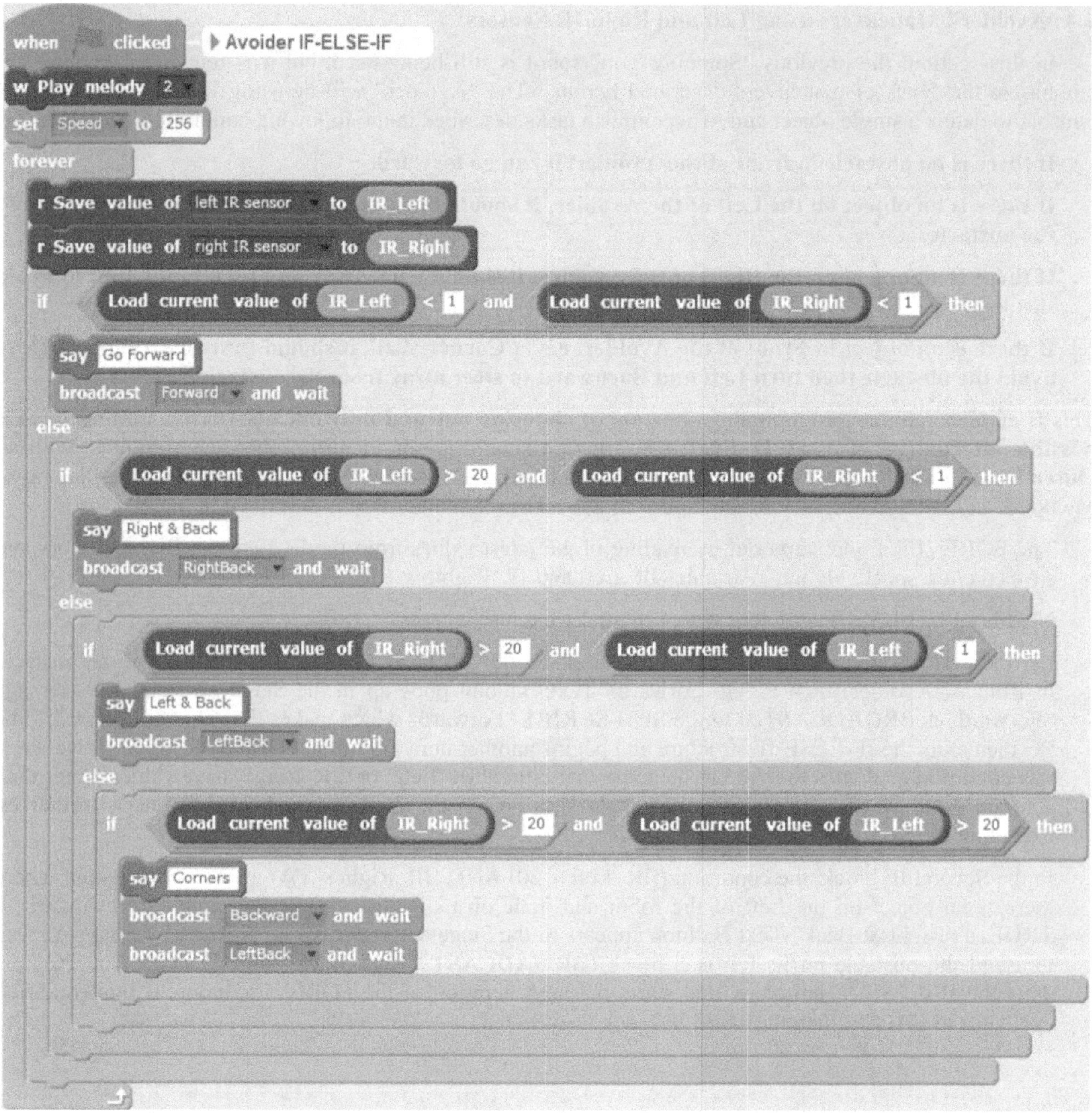

Fig. 4.13 Main SCRIPT used in program "Avoider-2.sb2" (IT-ELSE-IF used).

Similarly, the condition ((IR_Right > 20) AND (IR_Left < 1)) used for the Third IF block represents a situation where there is an object on the Right of the robot and none on its Left. If this condition evaluates to TRUE, a "Left & Back" Text Balloon appears in the Stage and the robot Backs away and turns Left to avoid the obstacle on its Right (see block **BROADCAST "LeftBack" and Wait**). The PC next exits this IF-ELSE-IF structure, and starts a new iteration of the FOREVER loop. If this condition evaluates to FALSE, then the fourth IF block is examined next.

The Fourth IF block uses this condition ((IR_Left > 20) AND (IR_Right > 20)) to represent a situation when there is an object close enough or large enough in front of the robot such that both IR Sensors report data values larger than 20. If this condition is TRUE, the "Corners" Text Balloon appears on the Stage and the PC BROADCASTs the first message "Backward" & WAITs for SUB-SCRIPT "Backward" to execute, then it BROADCASTs the second message "LeftBack" & WAITs for SUB-SCRIPT "LeftBack" to execute. Thus, the robot would first move straight back and make a left-back turn to steer it away from any potential corner or even a cul-de-sac.

Video 4.9 shows that this Avoider-2 algorithm has a good performance, but is less responsive than its TASK counterpart ("Avoider-2.tskx" and Video 3.12).

The next SCRATCH project "Avoider-3.sb2" offers a different solution technique to these maneuvering tasks. **This new approach still relies on the same condition/action listing as described at the beginning of this section 4.4, but it distills different behavior patterns from this list:**

1) We must identify a "default" action which in this case is "Going Forward".

2) We will be using several simple IFs in parallel instead of a single big IF-ELSE-IF structure like before. **We also need to do a deeper analysis of the condition/action list to retain only the "essential" condition/action pairs.**

3) We also **need to apply a Delay Time to each action made** (this part was not needed when the IF-ELSE-IF structure was used). This Delay Time can be tuned for optimal performance.

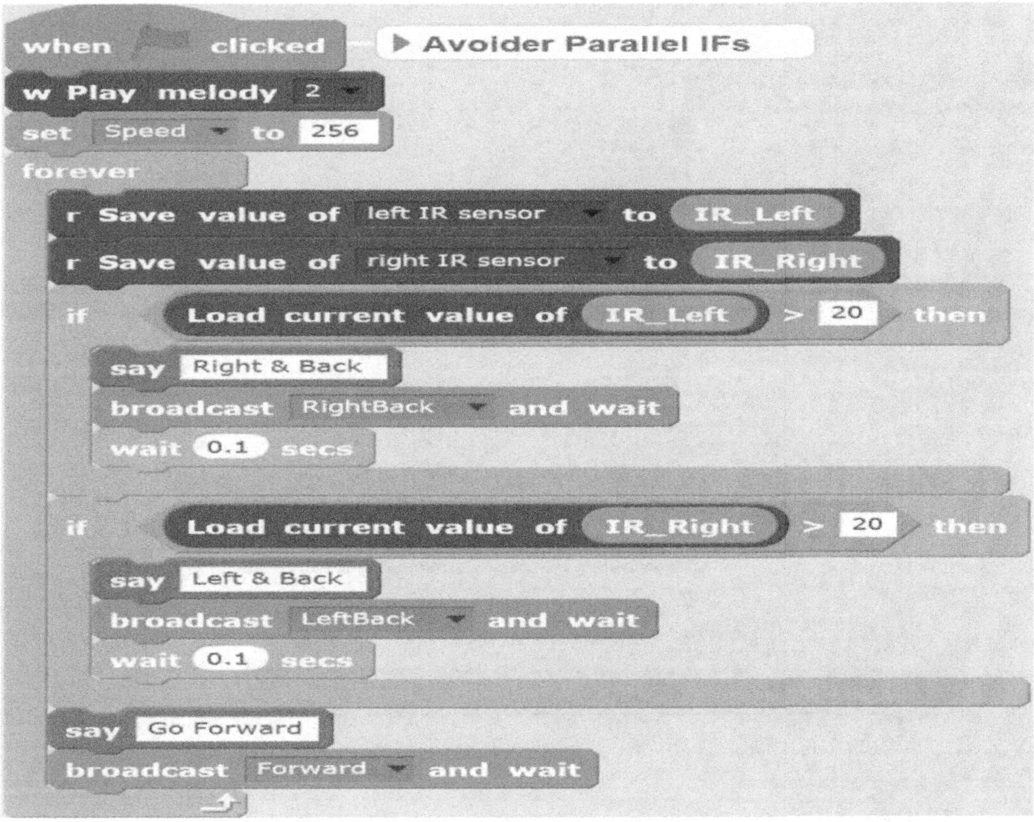

Fig. 4.14 Main SCRIPT used in program "Avoider-3.sb2" (parallel IFs used).

Fig. 4.14 shows the revised FOREVER loop:

At the start of the FOREVER LOOP, Variables IR_Left and IR_Right are updated with the latest data input from the IR Left and Right Sensors. Next are two parallel IF blocks.

The first IF block checks only on the condition for an object on the robot's Left (conditional expression "IR_Left > 20"). If this is TRUE, a "BROADCAST & WAIT" block is used to make the robot back up and turn right to steer itself away from the obstacle. A short delay of 0.1 second is necessary for the robot to stay in this "action" to make this algorithm work properly. The reader can try different values for the time delay to find the optimal robot performance.

The second IF block represents the condition when there is an object on the robot's right, then the robot backs up and turns left to steer itself clear from the obstacle, using another "BROADCAST & WAIT" block with the message "LeftBack". A short delay 0.1 second is also necessary for the robot to stay in this "action" to make this algorithm work properly.

Please note that the "default" maneuver is always programmed as the "Last Action" ("Going Forward" in this case).

The reader can readily see that this SCRATCH version is more compact than the "Avoider-2.sb2" version. Video 4.10 shows the runtime performance of this algorithm.

The next project "Avoider-4.sb2" offers a third approach to the same condition/action list shown at the beginning of Section 4.4, but now using "REPEAT UNTIL" loops (see Fig. 4.15). REPEAT UNTIL Loops are available in the CONTROL menu of the SCRIPT tab.

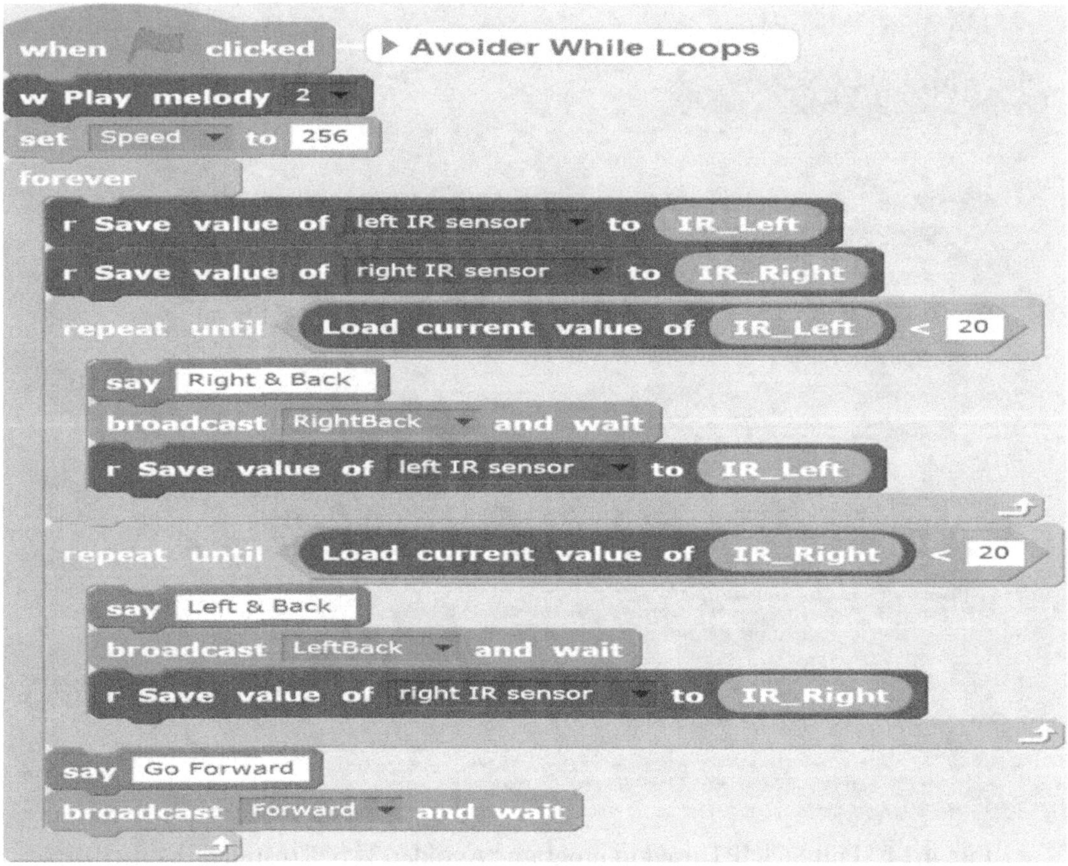

Fig. 4.15 Main SCRIPT used in program "Avoider-4.sb2" (parallel REPEAT UNTIL loops).

Fig. 4.15 shows that it also uses the "Default Maneuver" approach (same as "Avoider-3.sb2"), and the parallel REPEAT UNTIL loops match up with the parallel IFs (location wise). But there are three differences:

1) Delay Times are not needed for REPEAT UNTIL loops.

2) The user needs to make sure to read in the most recent data values for all Variables set up in the Conditional Expressions used by the REPEAT UNTIL loops, both before entering the loops for the first time and as the last action of any loop.

3) The user also needs to note that the Conditional Expression used in a REPEAT UNTIL loop is the Negation of an Equivalent Conditional Expression used in an IF block (please review and contrast Figs. 4.14 and 4.15).

Video 4.11 shows the runtime performance of the program "Avoider-4.sb2".

Does the reader have a preference over one of these 3 versions of Avoider? The big IF-ELSE-IF structure of "Avoider-2.sb2" may be too cumbersome. "Avoider-3.sb2" has a "tunable" feature in terms of the Delay Time. "Avoider-4.sb2" requires more executions of the "read" type Extension Command which costs 70-100 ms each time.

There is an interesting experiment that the reader can make with the previous different versions of the Avoider codes: create a maze using books or boxes and let the Avoider robot roam freely through it. You will be surprised with its behaviors depending on which SCRATCH code you used, especially when your maze has a cul-de-sac.

4.5 "Follower" Maneuvers using Left and Right IR Sensors

To round out the reader's experience with using the Left and Right IR Sensors, let's apply a **"reverse logic"** on them, essentially let's make the robot follow an object when it detects something in front of it.

Let's first learn how to track a moving object - when it is detected. The corresponding condition/action list can be as follows:

If there is no object detected, the robot should stay still (default behavior).

If there is an object detected by the Right IR Sensor, the robot should swing right.

If there is an object detected by the Left IR Sensor, the robot should swing left.

Should we also plan for the case when an object is detected and located right in front of the robot? Should it "stop" or should it show that it already "locks in" on the object in some other way? From the preceding Avoider SCRATCH coding work, we should be able to let the robot bounce between its two actions of swinging left and right to show that it got the object in its sight, right?

The program "Follower-1.sb2" was derived from the previous discussions using REPEAT UNTIL Loops (i.e. "Avoider-4.sb2"). The algorithm used is shown in Fig. 4.16:

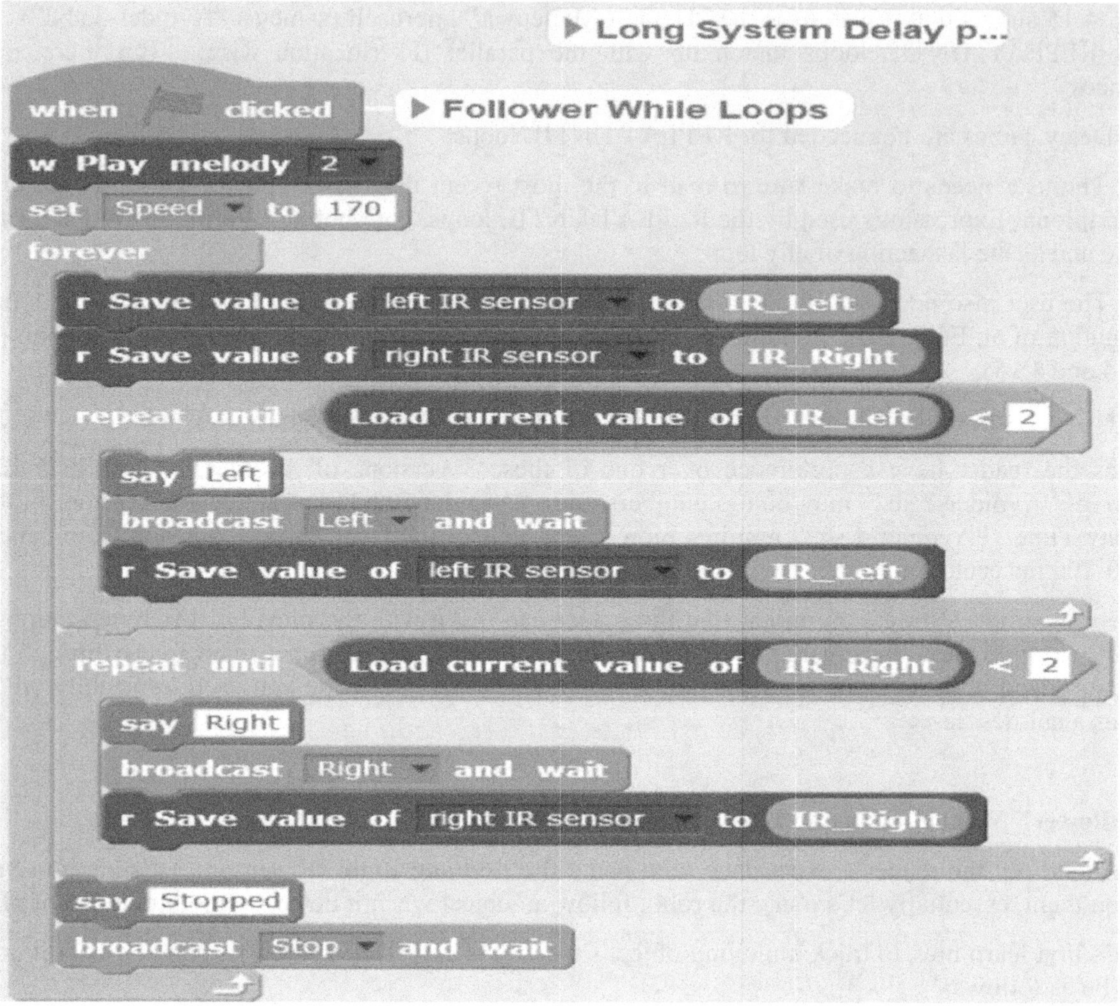

Fig. 4.16 Main SCRIPT used in program "Follower-1.sb2" (parallel REPEAT UNTIL loops).

Essentially, whenever any IR Sensor (Left or Right) picks up on an object, the robot rotates respectively towards the same side where that IR Sensor is located. These actions are implemented by the two REPEAT UNTIL Loops. Please note that a low detection threshold value of "2" is used so that the robot can detect an object as "far" as possible!

The default "Action/Maneuver" is to stop and it is used at the bottom of the FOREVER Loop.

Also, please note the structural similarity between "Avoider-4.sb2" and "Follower-1.sb2", except for the actual Conditional Expressions and Actions used, of course.

Video 4.12 shows the runtime performance of this algorithm. **Please note the longer system delay shown in the robot behavior and the effect of the object's size on how the robot would tract the object.**

Fig 4.17 shows another algorithm using parallel IFs instead, but yielding the same object tracking result. Please note that a small Delay Time is no longer needed as compared to an equivalent TASK code, as the SCRATCH/R+SCRATCH tool chain has plenty of system delay times to contend with already. Please also note the "reverse logic" in the Conditional Expressions used between "Follower-1.sb2" and "Follower-1-IFs.sb2". See Video 4.13 for a runtime performance of the program "Follower-1-IFs.sb2".

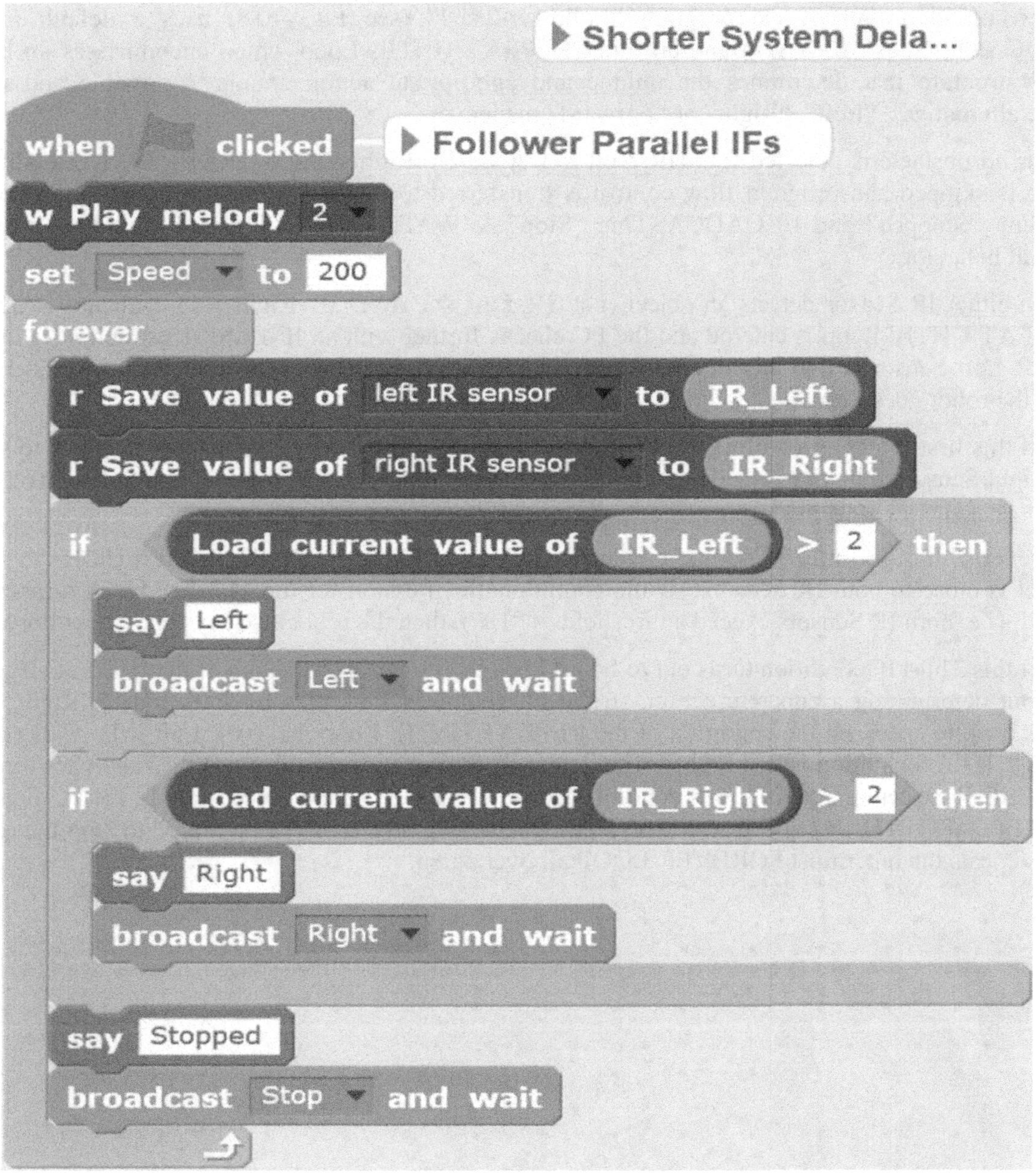

Fig. 4.17 Main SCRIPT used in program "Follower-1-IFs.sb2" (parallel IF blocks).

In the next "Follower" application, a new condition/action pair is added to the previous list:

When the object gets a bit closer to the robot (i.e. threshold value of 10 applied to both IR Sensors), the robot should get even closer to the object by going Forward. At this point during runtime, if the user then moves the object away from the robot at a proper speed, ones would see the effect of the robot "following" the object (at a short distance away).

There are two possible solutions: one using a **combination of REPEAT UNTIL and IF-ELSE-IF** structures (Follower-2.sb2), and another one using **only an IF-ELSE-IF** structure (Follower-3.sb2).

The Main algorithm of the program "Follower-2.sb2" (see Fig. 4.18) uses a default maneuver (BROADCAST "Stop" & WAIT) and an outer REPEAT UNTIL Loop which encompasses an inner IF-ELSE-IF structure that determines the unique and appropriate action among 3 possible and mutually exclusive alternatives: "Left", "Right" or "Forward" maneuvers:

When no obstacle is detected (i.e. (IR_Left < 1 AND IR_Right < 1) is TRUE), the REPEAT UNTIL Loop is skipped and program flow control is transferred to the bottom of the FOREVER LOOP, i.e. SAYing "Stopped" and BROADCASTing "Stop" & WAIT. Thus, the robot would stay still at its default behavior.

When either IR Sensor detects an object (i.e. (IR_Left < 1 AND IR_Right < 1) evaluates FALSE), the REPEAT UNTIL Loop is entered and the PC checks further with an IF-ELSE-IF structure to first see if the IR Left Sensor is the only one detecting the object (e.g. threshold "5" exceeded in First IF block) then the robot goes Left.

But if this first IF condition turns out to be FALSE, the PC next checks the Second IF block to see if the IR Right Sensor is the only one detecting the object (e.g. threshold "5" exceeded in Second IF), then the robot says and goes "Right".

However, if this Second IF Condition evaluates to be FALSE also, then the PC next checks to see if the object is close to both IR Sensors. If the conditional expression for the Third IF block turns out to be TRUE (i.e. both IR Sensors exceed a threshold of "10"), then the robot also says and goes "Forward".

But if this Third IF condition turns out to be FALSE, the PC reaches the end of the IF-ELSE-IF structure without determining a concrete action, so the PC refreshes Variables IR_Left and IR_Right and then loops back to check on the condition of the REPEAT UNTIL Loop, i.e. ((IR_Left < 1) AND (IR_Right < 1))". If this condition is still FALSE (i.e. either IR Sensor sees an object), it then searches through the IF-ELSE-IF structure again. If this condition becomes TRUE (i.e. no object found), the PC finally quits the REPEAT UNTIL Loop. It next SAYs "Stopped" and BROADCASTs "Stop" to stop the robot and then repeats the outermost FOREVER LOOP all over again.

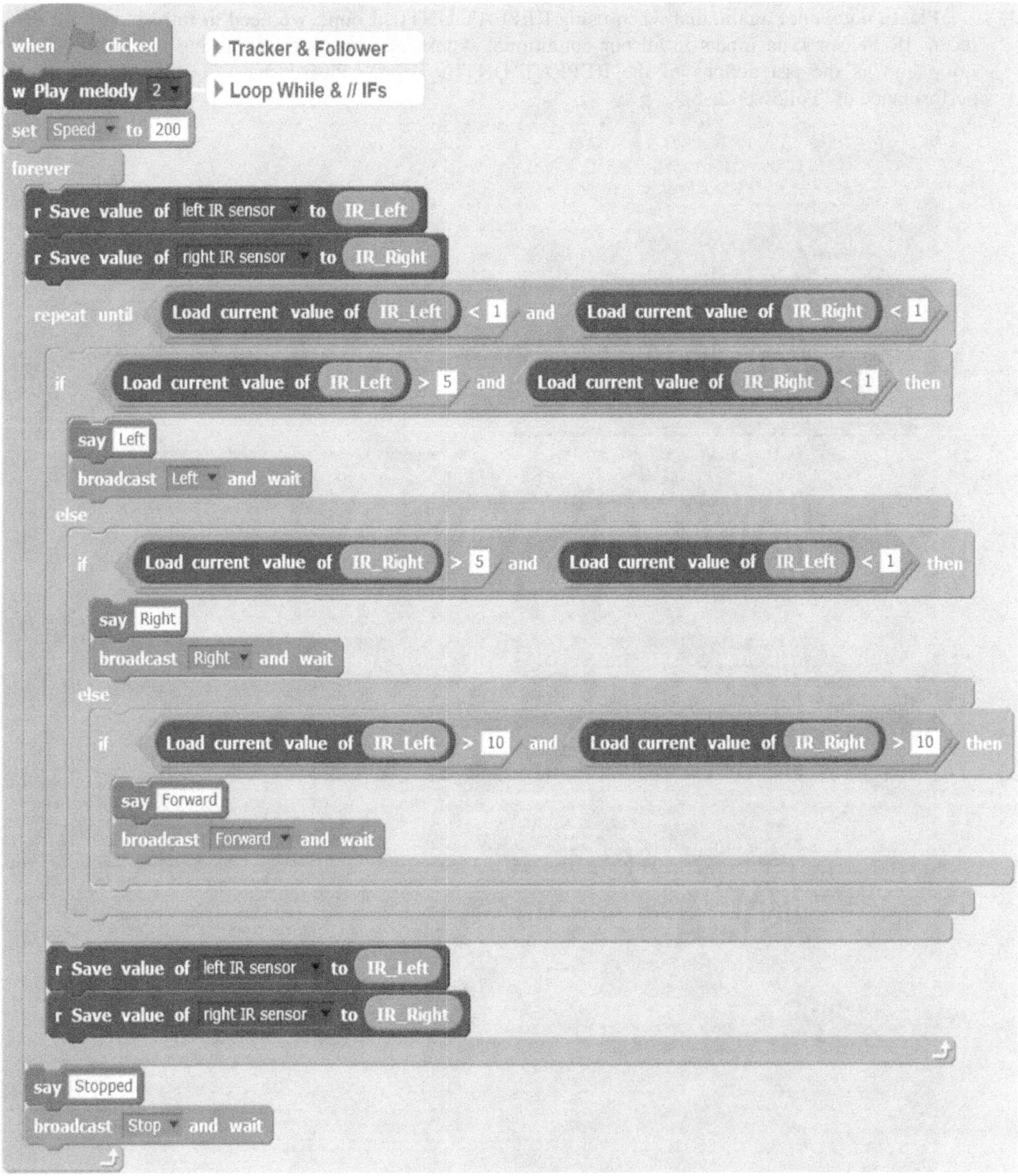

Fig. 4.18 Main SCRIPT used in program "Follower-2.sb2"

(Combination of LOOP and IF blocks).

Please note, once again, that when using REPEAT UNTIL Loops, we need to refresh and use the most recent IR Sensor Data inputs in all our conditional statements, just before entering the REPEAT UNTIL Loop and as the last actions of the REPEAT UNTIL Loop. Please see Video 4.14 for the runtime performance of "Follower-2.sb2".

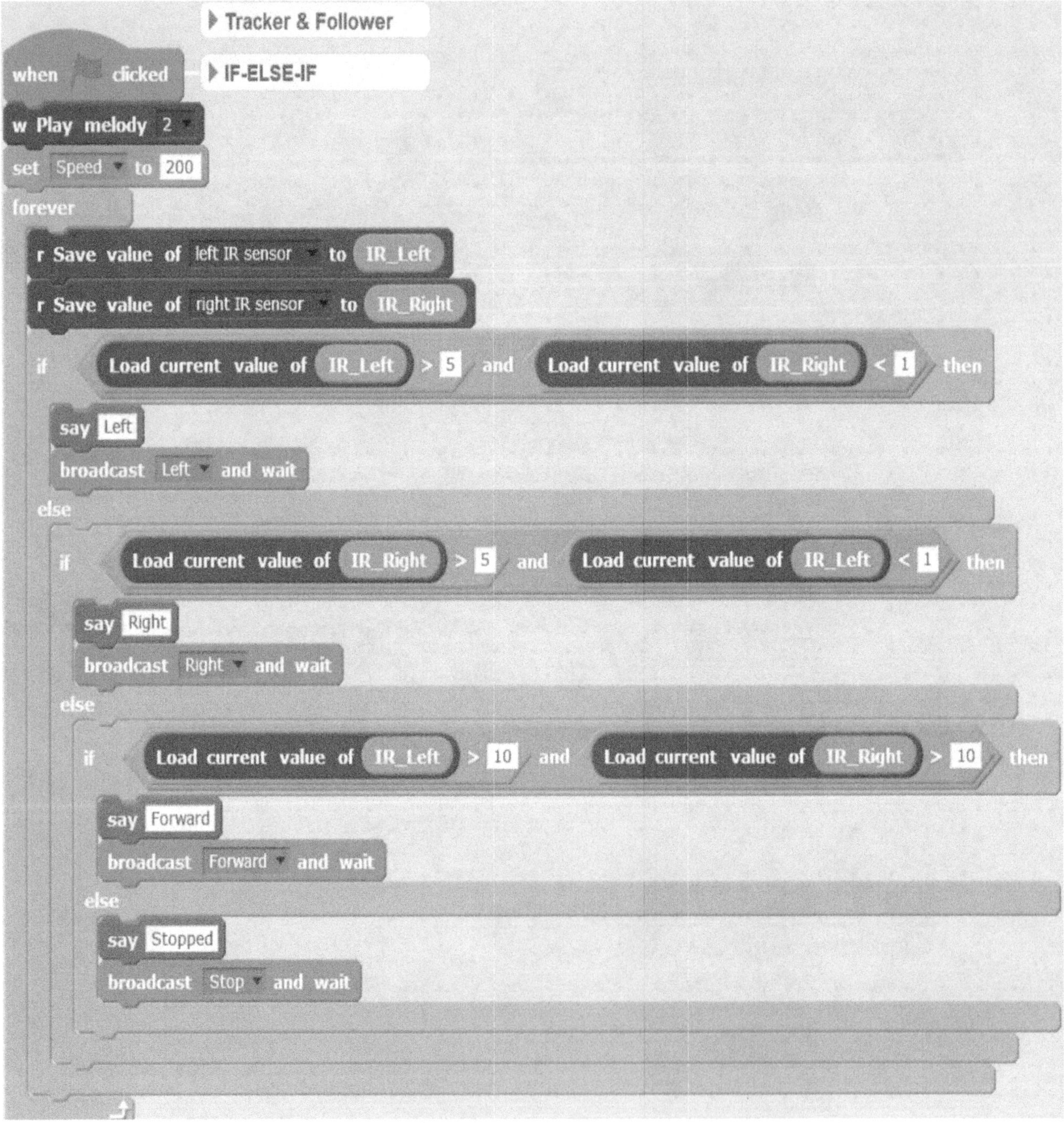

Fig. 4.19 Main SCRIPT used in program "Follower-3.sb2"

The algorithm used in the next program "Follower-3.sb2" is described in Fig 4.19. It is more straight forward so it is easier to explain:

It uses an outermost FOREVER LOOP (as always) and the first actions by the PC is to go out to the IR Sensors Ports and to read in the latest IR Sensor values and save them in Variables IR_Left and IR_Right.

Next is the familiar IF-ELSE-IF structure which now represents 4 mutually exclusive condition/action pairs for "Left", "Right", "Forward" or "Stop". Please note that the IF-ELSE-IF structure in the previous "Follower-2.sb2" handles only 3 condition/action pairs.

If the First IF is TRUE, i.e. only the IR Left Sensor sees the object, then the robot turns left.

Otherwise, and if the Second IF turns out to be TRUE, i.e. only the IR Right Sensor sees the object, then the robot turns right.

Otherwise, and if the Third IF turns out to be TRUE, i.e. both IR sensors see the object up close, then the robot goes forward to get even closer to the object.

Lastly, when no condition turns out to be TRUE (i.e. the last ELSE block), the robot executes its default behavior, i.e. staying still.

Video 4.15 shows the run-time performance of this program "Follower-3.sb2".

There was no discernable difference in performance between these two solutions that the author could see. In theory, "Follower-3.sb2" should be "faster" because it accesses the sensor ports only twice per each iteration of the overall FOREVER LOOP, i.e. it uses less resources from the SCRATCH/R+SCRATCH tool chain as compared to "Follower-2.sb2" which uses from 2 to 4 accesses to the sensor ports per each iteration of its overall FOREVER LOOP.

Recapping Sections 4.4 and 4.5, the author's goal is to show that programming "autonomous" behaviors is fun and challenging at the same time! Starting from the (now familiar) condition/action analysis, **three techniques can be used to create SCRATCH codes and after some initial actions, they all use an outermost FOREVER LOOP for the main algorithm:**

1) In Technique 1, for each iteration of the FOREVER LOOP, latest readings from the used sensors are first recorded, then an IF-ELSE-IF structure is used to represent all possible but mutually exclusive condition/action pairs that correspond to robot behaviors required to fulfill a given challenge. Example programs using Technique 1 are "IR-based-Maneuvers.sb2", "IR-based-Maneuvers-Say.sb2", "Avoider-2.sb2", and "Follower-3.sb2".

2) In Technique 2, the sensors reading task is still required first, but the IF-ELSE-IF structure is replaced by "parallel" IFs. This technique requires a deeper analysis of the original condition/action list to identify a "default" behavior, and other condition/action pairs that can work "on their own" or "in concert". This technique also may require a short time delay incorporated into each robot action. Compared to Technique 1, Technique 2 usually results in a smaller number of statements used in the SCRATCH code, meaning that the PC should take less time to do each iteration of the FOREVER LOOP, and thus the robot may become more responsive to changes in the "World" (review Fig. 1.10). Example programs using this technique are "Avoider-3.sb2" and "Follower-1-IFs.sb2".

3) Technique 3 is a variation of Technique 2 and it uses "parallel" REPEAT UNTIL Loops instead of "parallel" IFs. The user would need to **coordinate** "Data Acquisition" tasks in its Conditional Statements and at appropriate parts of the REPEAT UNTIL Loops. Thus, the user should expect more system delays as compared to the other two Techniques, as "Data Acquisition" commands cost 70-100

ms each to go through the SCRATCH/R+SCRATCH/CM-50 hardware/software chain. Example programs using Technique 3 are "Avoider-4.sb2" and "Follower-1.sb2". "Follower-2.sb2" is a hybrid solution combining Techniques 3 and 1.

4.6 Remote Control of CarBot via PC's keyboard

The SCRATCH/R+SCRATCH tool chain does not support a Virtual RC-100 Remote Controller like for the TASK tool (see Section 3.6), as SCRATCH already has a block that monitors the PC's keyboard, called "When xxx Key Pressed" and it is available in the EVENTS menu of the SCRIPTS tab (see Fig. 4.20).

Fig. 4.20 EVENTS Menu in SCRIPTS Tab

Another big difference between the RC-100's behavior and the PC's keyboard's behavior is in the way multiple buttons or keystrokes are handled. At any one time, the RC-100 sends over to the robot a 10-bit message that represents the status (OFF or ON) of all its 10 buttons (UDLR123456), thus the user needs to create special TASK codes to separate specific data from each button before ones can use them in a remote-control scheme (see Section 3.6). On the other hand, when multiple key strokes are made on the PC keyboard, the SCRATCH IDE receives them as a stream of individual keyboard inputs, i.e. none of the communications processing techniques created for TASK in Section 3.6 is transferrable to SCRATCH.

In SCRATCH, the general approach is to create independent SUB-SCRIPTS to handle each key stroke of interest, and a MAIN SCRIPT to handle the interactions between keystrokes. The project "RC-Direction-Speed-1Button-a.sb2" shows how to use the Up-Down-Left-Right Arrow keys for affecting the moving directions of a typical CarBot (see Fig. 3.55) and the keys "s" and "f" to switch respectively between a Slow Speed and a Fast Speed regime.

Fig. 4.21 shows only the SUB-SCRIPT handling the "Up-Arrow Key" when it is sent from the PC into the SCRATCH IDE (the other Arrow keys are handled similarly):

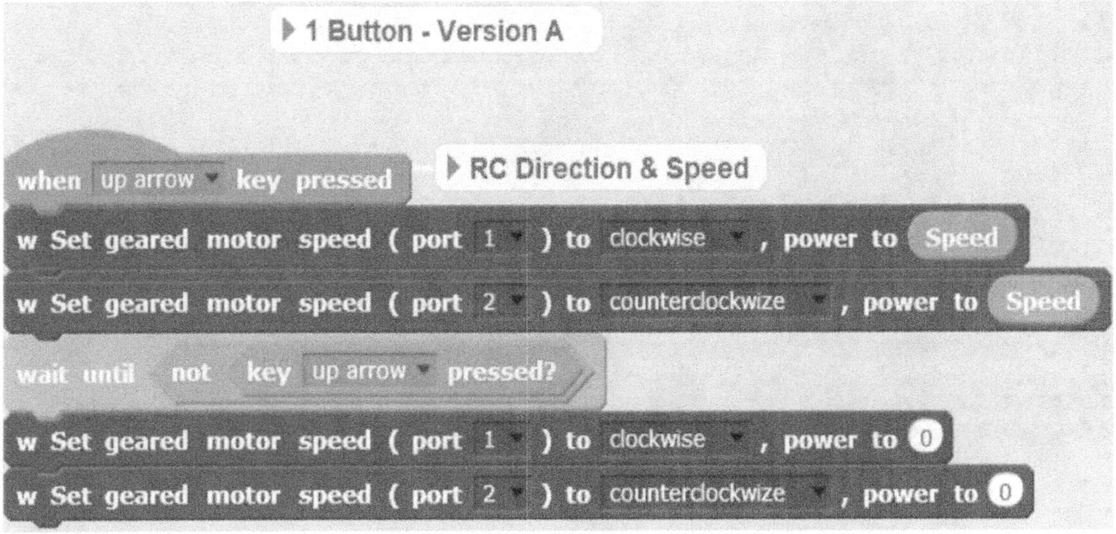

Fig. 4.21 SUB-SCRIPT handling the Up-Arrow Key.

When the Event "Up-Arrow Pressed" is detected, Ports 1 and 2 are set to a Power Level equal to the value of Variable Speed, with appropriate motor turn directions to make the CarBot go forward.

Next a WAIT UNTIL block (from the CONTROL Menu) is used to monitor when the user releases the Up-Arrow Key, i.e. when the Up-Arrow key is NOT PRESSED.

The last actions are to turn off power on the two motors to put the CarBot into a STOP.

Fig. 4.22 shows how the MAIN SCRIPT can get started to set Variable Speed to 500, and how the "f" key is monitored to set Speed to 800 and how the "s" key is monitored to set Speed to 250.

Fig. 4.22 MAIN SCRIPT and SUB-SCRIPTs handling the "s" and "f" Keys.

Video 4.16 shows how this scheme ("RC-Direction-Speed-1Button-a.sb2") works out as a Remote-Control scheme whereas only ONE button can be processed at any one time, although the user can push down multiple keys at the same time. Video 4.16 shows a very responsive RC system for 1 button to be used at a time.

The project "RC-Direction-Speed-1Button-b.sb2" shows an alternate solution where the job for handling the release of the Arrow Keys is shifted into the MAIN SCRIPT (see Fig. 4.23):

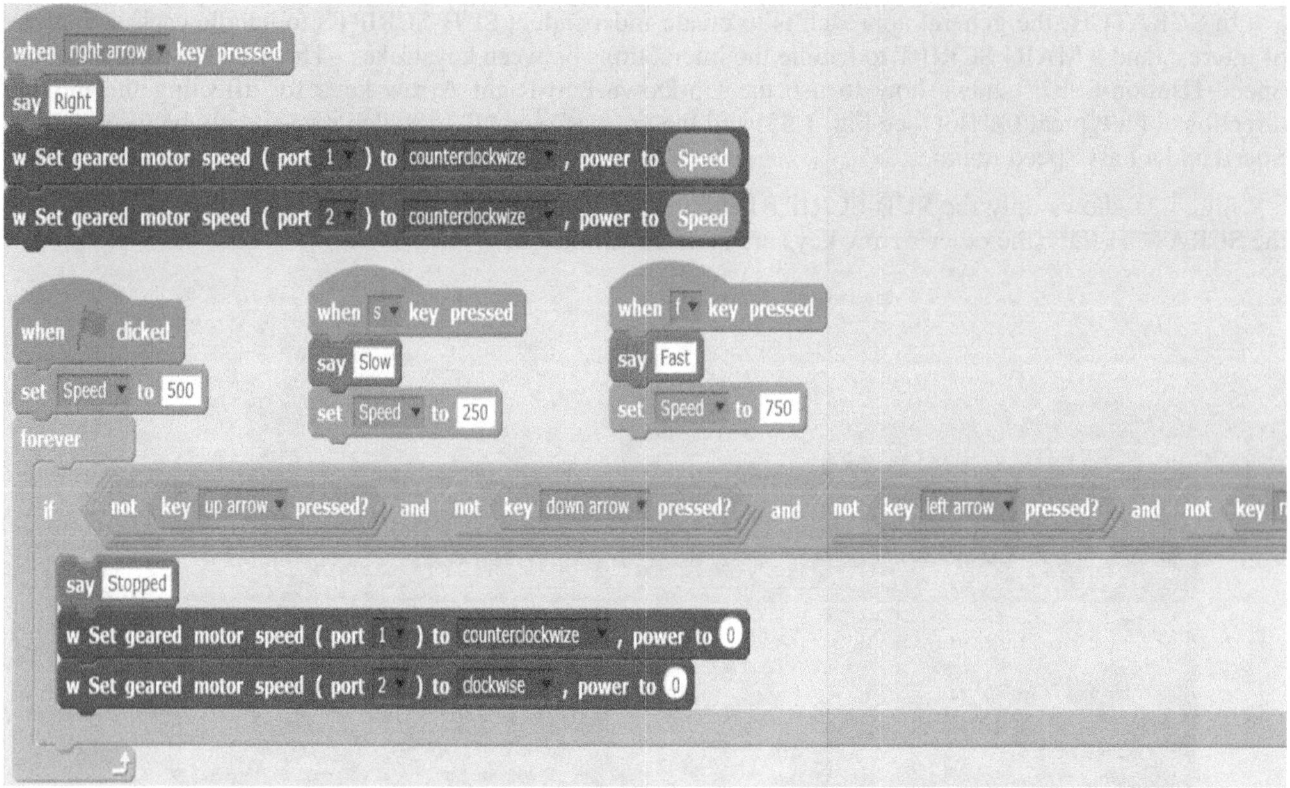

Fig. 4.23 MAIN SCRIPT and selected SUB-SCRIPTs in "RC-Direction-Speed-1Button-b.sb2".

The MAIN SCRIPT starts out by SETting Variable Speed to 500, then it enters an FOREVER LOOP. Inside this loop, an IF block, with a rather complex "AND" Conditional Expression representing the situation when NO Key Strokes are detected, is used to SAY "Stopped" and to turn power off on both motors.

Fig 4.23 also shows "enhanced" SUB-SCRIPTS for key strokes "s" and "f" using appropriate Text Balloons.

The SUB-SCRIPT for "Right Arrow" is typical for all Arrow Keys, and it shows that these SUB-SCRIPTs are now only concerned with setting appropriate turn directions and power levels to the motors to make the robot go in the user-wanted directions.

The project "RC-Direction-Speed-2Buttons.sb2" allows a combination of key strokes to be activated by the user: for example, "Up and Right" or "s and Left". This capability is achieved by putting Parallel IFs inside the MAIN FOREVER LOOP. Each IF block is designed to handle a given key stroke being pressed (see Fig. 4.24), including the case when no key is pressed (see Fig. 4.25).

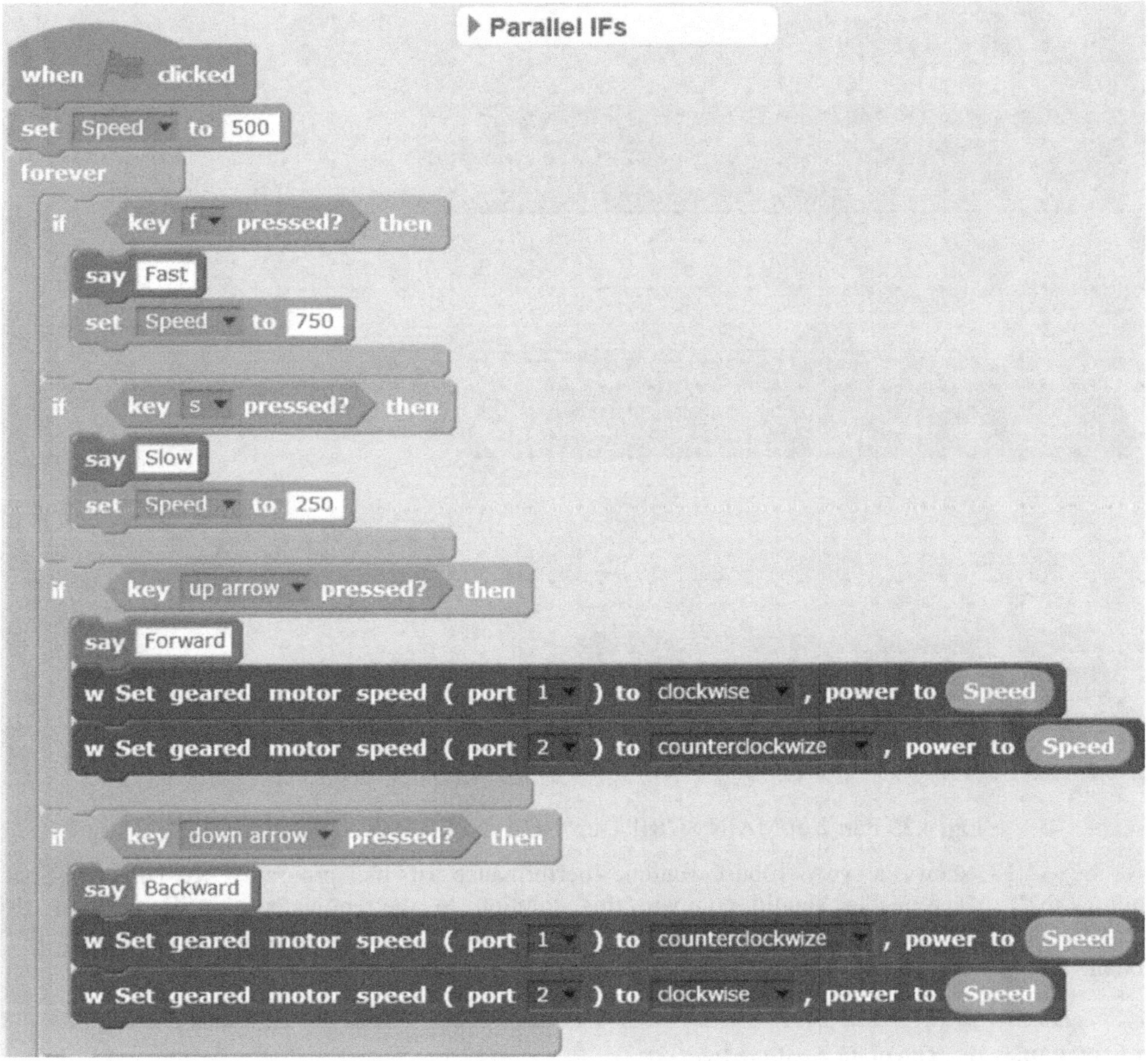

Fig. 4.24 Part 1 of MAIN SCRIPT in "RC-Direction-Speed-2Buttons.sb2".

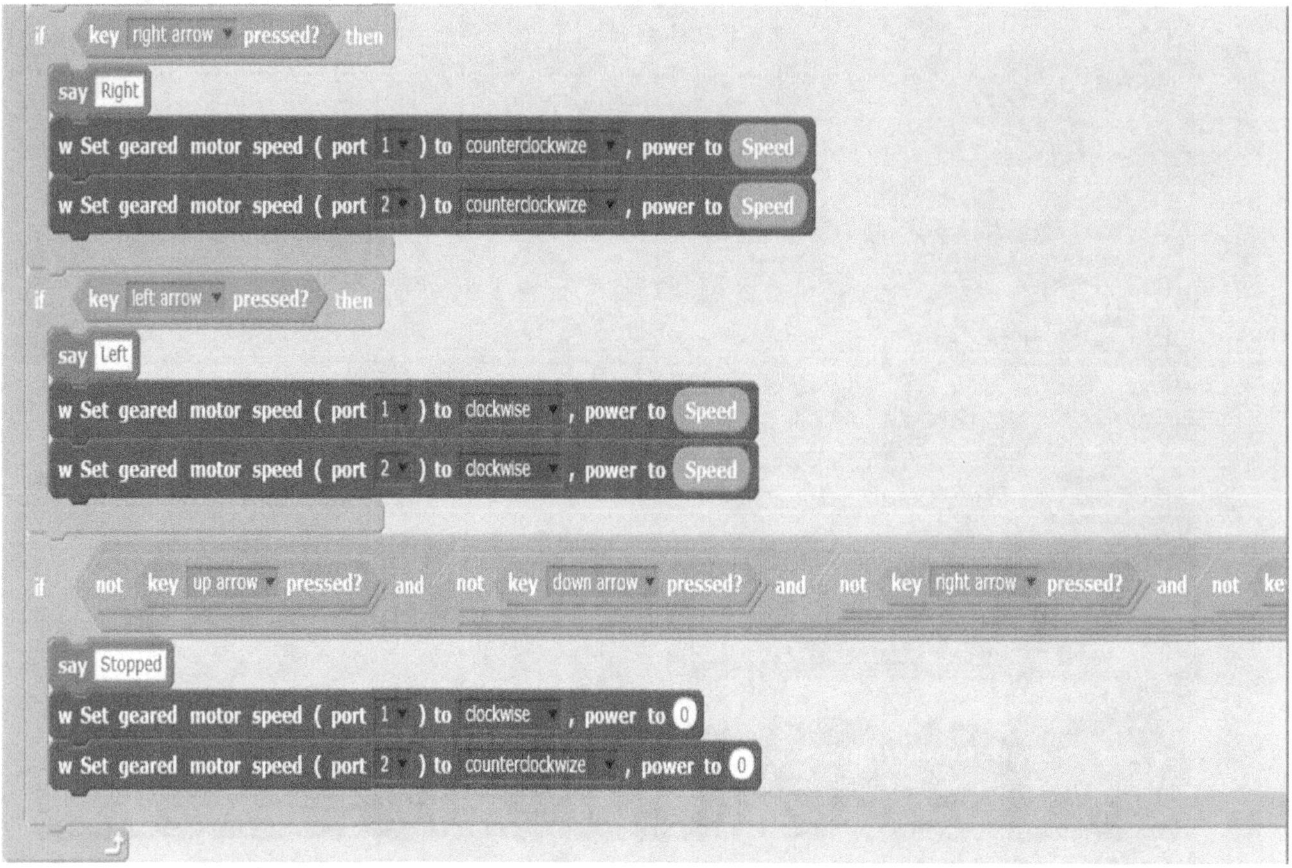

Fig. 4.25 Part 2 of MAIN SCRIPT in "RC-Direction-Speed-2Buttons.sb2".

Video 4.17 shows a very robust runtime performance of the program "RC-Direction-Speed-2Buttons.sb2". The reader should compare this solution to the equivalent TASK solution "RC-MultiDirectionsSpeeds.tskx" (see Section 3.6.2 and Video 3.21) – a lot more work was needed in the TASK version!

4.7 Line Follower – Obstacle Avoider Project

In this section, we'll develop a SCRATCH solution equivalent to the TASK program "LineFollower-ObstacleAvoider.tskx" from Section 3.7. It is recommended that the reader reviews Section 3.7 to recall the behaviors of the Left and Right IR Sensors and to understand the threshold values used in the program "LineFollower-ObstacleAvoider.sb2". The same algorithm, described in detail in Section 3.7 (which won't be repeated here), is used for the SCRATCH solution, of course with the syntax adapted to the equivalent blocks allowed in SCRATCH.

Fig. 4.26 describes the MAIN SCRIPT of "LineFollower-ObstacleAvoider.sb2":

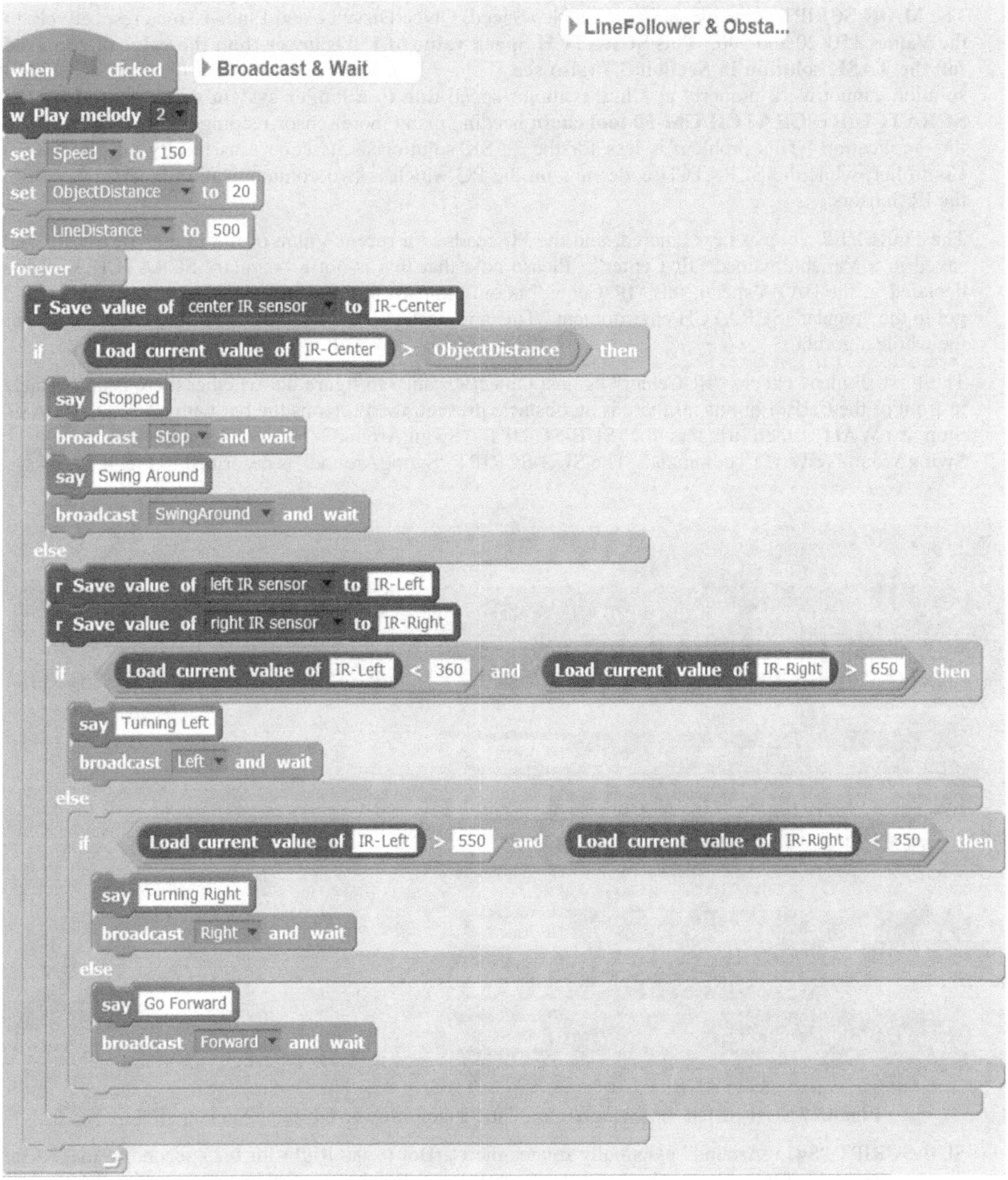

Fig. 4.26 MAIN SCRIPT in "LineFollower-ObstacleAvoider.sb2".

The MAIN SCRIPT starts by setting Variables Speed, ObjectDistance and LineDistance respectively to the Values 150, 20 and 500. **This SCRATCH Speed value of 150 is lower than the value of 256 used for the TASK solution in Section 3.7** (also see Video 3.27), because it turns out that the SCRATCH solution cannot work properly at a higher motor speed **due to a longer system delay caused by the SCRATCH/R+SCRATCH/CM-50 tool chain** needing many more sensor-reading cycles than before in this application. This problem is less for the TASK solution as its code runs locally on the CM-50 Controller, while the SCRATCH code runs on the PC which is two communication layers away from the IR Sensors.

The FOREVER Loop is next entered, and the PC reads in a recent Value of the Center IR Sensor and saved in a Variable named "IR-Center". Please note that this is not a "regular" SCRATCH Variable declared in the DATA menu, this "IR-Center" is only known to the ROBOTIS Extension Blocks, and not to the "regular" SCRATCH environment. This approach seemed to shave off a few milliseconds for the whole algorithm.

The First IF block checks "IR-Center" against ObjectDistance to figure out whether there is an obstacle in front of the CarBot or not. If there is an obstacle present, the PC stops the bot with a "BROADCAST Stop & WAIT", then triggers the SUB-SCRIPT "SwingAround" by issuing the "BROADCAST SwingAround & WAIT" command. The SUB-SCRIPT "SwingAround" is described in Fig. 4.27.

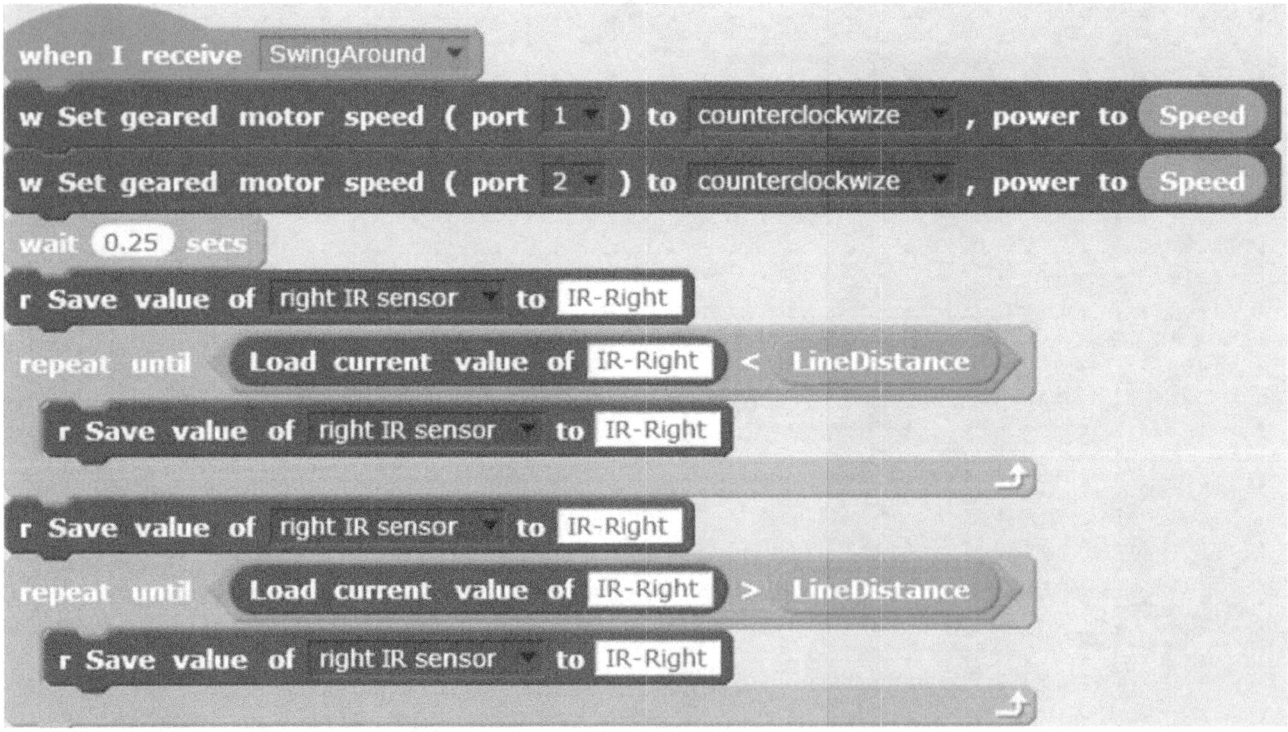

Fig. 4.27 SUB-SCRIPT "SwingAround" in "LineFollower-ObstacleAvoider.sb2".

SUB-SCRIPT "SwingAround" essentially rotates the CarBot to the Right for 0.25 second to make sure that the Right IR Sensor is off the black track and on the background paper surface (i.e. IR-Right's Value is now greater than LineDistance = 500). The First REPEAT UNTIL loop is designed to be executed as long as the most recent IR-Right reading is still greater than LineDistance, meaning that the

CarBot is still seeing the background paper surface, and therefore the "Rotate Right" motion should be maintained. Eventually, the CarBot crosses over the black line, and IR-Right's Value would be less than LineDistance, and the First REPEAT UNTIL loop is exited and the Second REPEAT UNTIL loop is taken up. At this point, the IR Right Sensor sees the black track, so IR-Right's Value is smaller than LineDistance, thus the Second REPEAT UNTIL loop should be maintained. But eventually the CarBot crosses over into the background paper area, and IR-Right's Value is now larger than LineDistance and the Second REPEAT UNTIL loop is exited. At that time, the PC program flow control pointer is passed back into the MAIN SCRIPT at the end of the First IF block, and the algorithm restarts the FOREVER LOOP.

If there is no frontal obstacle, the ELSE branch of the First IF would be taken up, whereas an IF-ELSE-IF structure is used to determine three mutually exclusive condition/action pairs:

IF IR-Left sees Black and IR-Right sees White (i.e. background paper surface), then the CarBot should rotate into the Left to align the CarBot again with the Black line.

ELSE, IF IR-Left sees White and IR-Right sees Black, then the CarBot should rotate into the Right to align the CarBot again with the Black line.

Otherwise, i.e. the Final ELSE, the CarBot should just drive Forward. At that time, the CarBot could be right on top of the Black Line or it could be right on top of the background paper (i.e. it is off-track). This uncertainty is unavoidable due to the placements of the Right and Left IR Sensors with respect to the width of the Black line (see previous discussions in Section 3.7).

Please see Video 4.18 for the runtime performance to this program "LineFollower-ObstacleAvoider.sb2".

4.8 Moody Dog Multimedia Project with SCRATCH

We will be using a 4-Legged DogBot for this project (see Fig. 4.28). It would be interesting for the reader to compare this SCRATCH version to the TASK/R+m.PLAY700 version in Section 3.9.

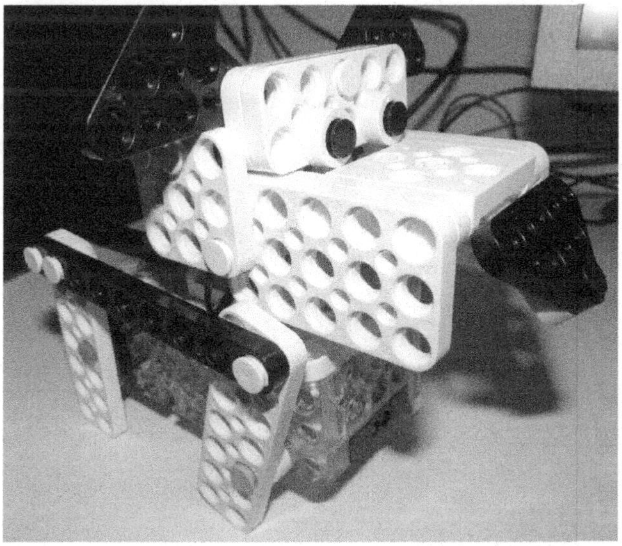

Fig. 4.28 4-Legged DogBot for Project "MoodyDogBot.sb2".

Fig. 4.29 illustrates the Costumes and Sound Clips used by the Moody DogBot SCRATCH project. One advantage when using SCRATCH is that this tool will tell its users how long would their audio clips last (i.e. around 5 seconds in our case).

Fig. 4.29 Costumes and Sound Clips used by Project "MoodyDogBot.sb2".

Similar to the TASK version, the SCRATCH project has a MAIN SCRIPT and two SUB-SCRIPTs, one for the Friendly mood, and another for the Angry mood.

Fig. 4.30 lists the blocks used in the MAIN SCRIPT:

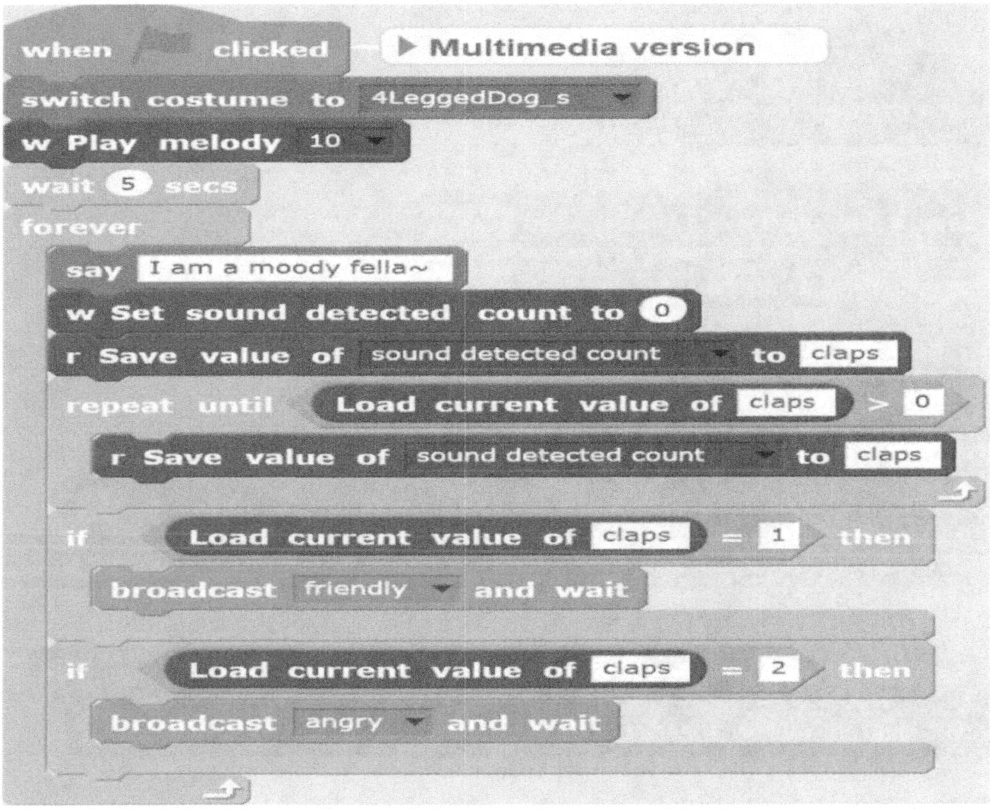

Fig. 4.30 MAIN SCRIPT in Project "MoodyDogBot.sb2".

Once the GREEN FLAG is clicked at runtime, the DogBot SPRITE switches to its 4 Legged Costume in the STAGE area, and the CM-50 is commanded to play Melody 10, then SCRATCH IDE waits for 5 seconds for Melody 10 to finish playing.

The FOREVER LOOP is entered and the DogBot announces that "I am a moody fella~".

Next SCRATCH sends out commands to the CM-50 to set its "Sound Detected Count" to zero, then reads in the current value of "Sound Detected Count" and saves it into Variable "claps". At this point, "claps" has a zero value, thus when the REPEAT UNTIL loop is entered next, the conditional expression (claps > 0) evaluates to FALSE, therefore this loop is entered. Next is a read command to the CM-50 to refresh the value of Variable "claps" and the conditional expression (claps > 0) is checked again. If the user stays real quiet, this REPEAT UNTIL would keep on looping. Please note that the structure, composed of an **initial** "Save Value Of" block, a "REPEAT UNTIL" **loop** and an **internal** "Save Value Of" block, is a very effective way to monitor a Sensor reading until a certain threshold is reached, causing the REPEAT UNTIL loop to quit. The reader will see this structure again and again.

When the user starts clapping, the REPEAT UNTIL Loop is exited, and the First of the two parallel IFs is processed to check on whether "claps" equals 1? If this is TRUE, a (BROADCAST "friendly" and WAIT) is launched to trigger the SUB-SCRIPT "friendly".

Fig. 4.31 illustrates the details of SUB-SCRIPT "friendly":

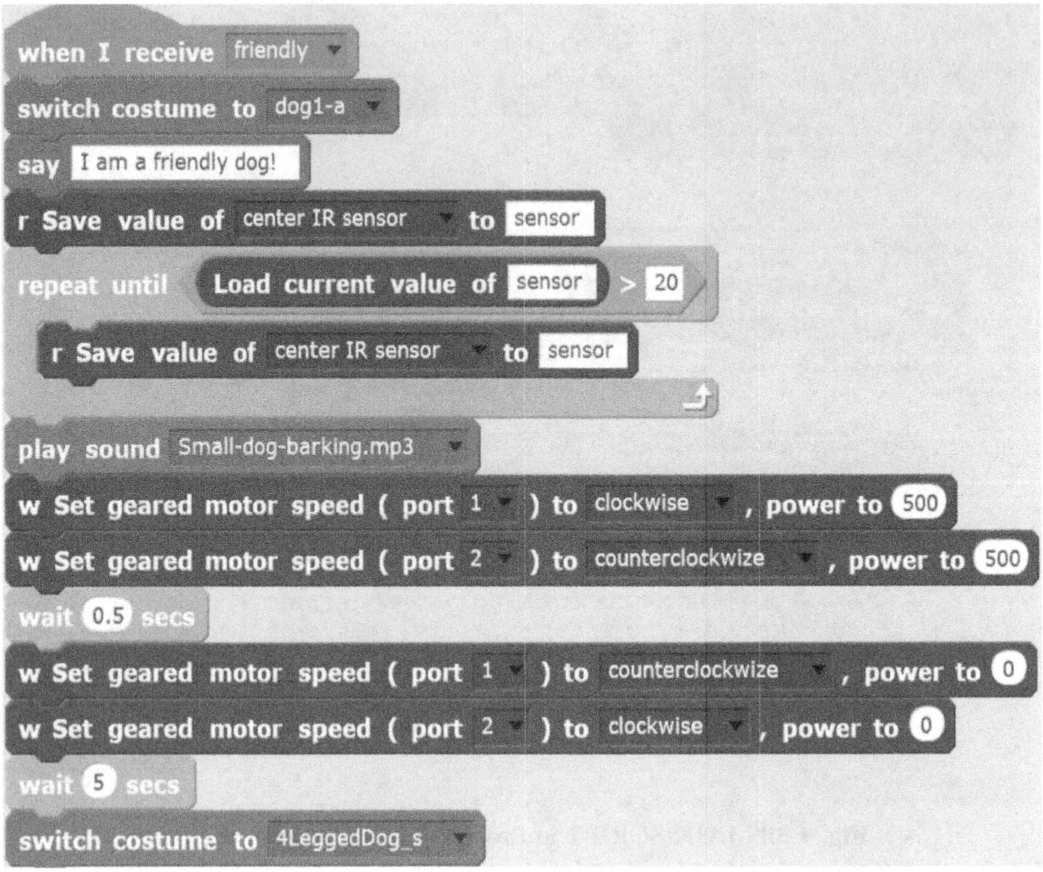

Fig. 4.31 SUB SCRIPT "friendly" in Project "MoodyDogBot.sb2".

When SUB-SCRIPT "friendly" is activated, the DogBot puts on new costume "dog1-a" and SAYs "I am a friendly dog!".

Then a Sensor Monitoring Loop is used again, but this time it is used to monitor the Center IR Sensor. It is meant to detect when there is an object close enough to the Center IR Sensor to cause it to read a Value greater than 20, at that time the REPEAT UNTIL loop is exited.

SCRATCH starts playing the audio clip "Small-dog-barking.mp3" (about 5.21 seconds long) on the PC. Next, it sends commands over to the CM-50 to make the DogBot move Forward at power level 500 for 0.5 second, and then turn off the motors. The "Wait 5 seconds" block is used to make sure that the audio clip "Small-dog-barking.mp3" has finished playing before the DogBot SPRITE switches back to the "4 Legged Dog" costume in the STAGE area.

At this point, the PC Flow Control pointer is transferred back to the MAIN SCRIPT and the second IF block is checked on but we know that it would evaluate to FALSE as "claps" is still set to 1. So, the FOREVER LOOP is picked up again for a fresh iteration.

And if in the next iteration of the FOREVER LOOP, the user happens to clap twice then we would witness the "Angry" SUB-SCRIPT unfold! See Fig. 4.32.

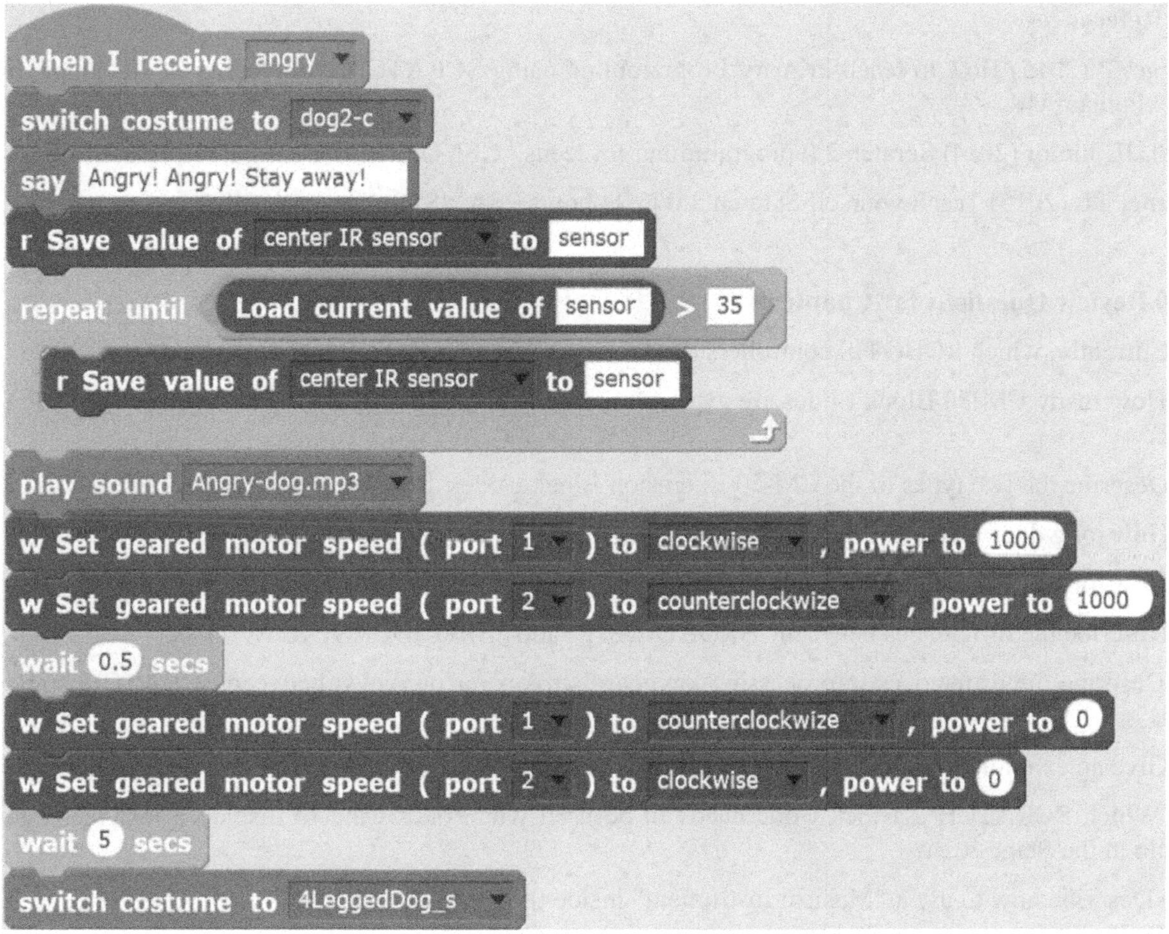

Fig. 4.32 SUB SCRIPT "angry" in Project "MoodyDogBot.sb2".

The reader may wonder what happens if the user claps more than twice? Well nothing happens to the DogBot because it is not yet programmed to respond to such a situation. The reader should feel very comfortable by now to expand this project to take care of such a situation!

Please see Video 4.19 for a runtime performance of this Moody DogBot SCRATCH project.

Recapping Chapters 1 through 4, we can draw some generalized conclusions:

1) The "Sense-Think-Act" paradigm and the condition/action analysis procedure had been helpful in the beginning steps of understanding a given robotic challenge and to sort out the relevant condition/action pairs.

2) At the next steps, the actual software tools (TASK or SCRATCH) we use to solve this challenge may produce different runtime performances on the same physical robot, and may even lead us to have to change the original goals of the challenge.

3) If the robotic challenge requires rather quick actuator responses or fast changing environmental inputs, then TASK is a more suitable tool to use.

4.9 References

Bagge, P (2015) How to teach Primary Programming using SCRATCH. University of Buckingham Press, Buckingham UK.

Ford, JL Junior (2014) Scratch 2.0 programming for teens. CENGAGE Learning PTR, Boston USA.

Warner TL (2015) Teach yourself Scratch 2.0 in 24 hours. SAMS, Indianapolis USA.

4.10 Review Questions for Chapter 4

1. Currently, which ROBOTIS controllers does the software tool R+SCRATCH support?

2. How many CM-50 Block Codes are available in the ROBOTIS Extension Package for the SCRATCH 2 IDE?

3. Describe the two types of the CM-50 Extension Block Codes available in the SCRATCH 2 IDE.

4. How long does it take for R+SCRATCH to process a "r" CM-50 Block Code?

5. How long does it take for R+SCRATCH to process a "w" CM-50 Block Code?

6. Describe the differences between "BROADCAST" and "BROADCAST & WAIT".

7. Describe the differences in processing keyboard strokes (arrow keys) between SCRATCH 2 and TASK tools.

8. Give an example when the ROBOTIS's "Load Current Value Of" block code has to be used.

9. Which SCRATCH 2 Block Code needs to be used when ones want to display a text message by the Sprite in the Stage Area?

10. Describe how to use a "Musical Instrument" inside the SCRATCH 2 IDE.

11. What is the critical step when using the REPEAT UNTIL loop structure inside the SCRATCH 2 IDE?

Chapter 5: Selected Mechanical Design Concepts

Although the parts in the PLAY700 kit are made of plastic, they are designed to mimic standard mechanical parts found in everyday life. Thus, there are many sources of information on mechanical engineering and design that the readers can draw from.

ROBOTIS has its own materials but some of their web materials are still in Korean, thus the reader will need to use Chrome as web browser so that it can translate them into "useable" English:

1) The OLLO K set is a more expanded version of the PLAY700 kit and it is only available in Korea at present, but the author has used it as an inspiration source for many mechanical designs described in this chapter (http://www.robotis.com/index/product.php?cate_code=10121111).

2) At present, the R+DESIGN tool has 3D CAD mechanical designs for up to the PLAY600 kit only, but the author is confident that the materials for the PLAY700 kit will be incorporated into this tool in future. R+DESIGN can be downloaded at (http://www.robotis.us/roboplus2/) and its user manual can be viewed at (http://support.robotis.com/en/software/roboplus2/r+design/r+design.htm). R+DESIGN can be used to design and assemble your own robot design but that will have to be the topic of another future book for the author!

Cornell University has a comprehensive library of mechanisms at its KMODDL web site (http://kmoddl.library.cornell.edu/). For example, if the reader wants to know more about the 4-bar linkage used for the Scorpion robot, this web page is quite comprehensive (http://kmoddl.library.cornell.edu/).

There are other numerous web sites for mechanism design and simulation, the author is listing just a few below:

http://www.mekanizmalar.com/

https://www.youtube.com/user/thang010146/videos

https://www.youtube.com/playlist?list=PLhoXNQqrCmEfAaTf0AfQ1Ztxmz2DoZiCk

http://www.roboanalyzer.com/mechanalyzer.html

5.1 Wheel Based Applications

5.1.1 Passive Front Supports/Wheels

As a typical PLAY700 CarBot has only two drive wheels, at the minimum it would need some Single Bottom Support (see Fig. 5.1). For smooth hard surfaces, these are OK to use, but they would dig into softer surfaces like carpet or sand and dirt.

Fig. 5.1 Two types of Single Bottom Support for CarBot.

The small White Washers can also be used as Side Supports (see Fig. 5.2), as these washers can "freely" roll, they would reduce friction with the ground surface and thus the CarBot can be more maneuverable.

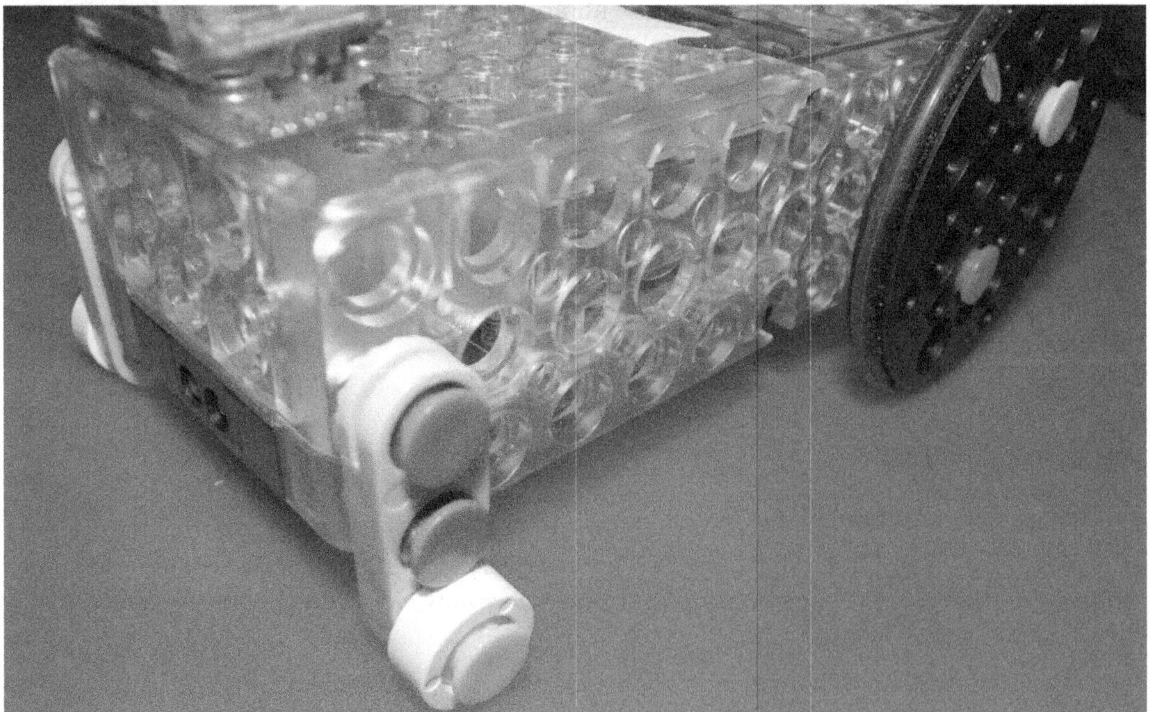

Fig. 5.2 White Washers as Front Wheels.

A better Front Wheel solution is to use the Small Black Pulley Wheels (see Fig. 5.3), without or with the small rubber tire (from the PLAY600 kit).

Fig. 5.3 Small Black Pulley Wheel as Front Wheels.

5.1.2 Four-Wheel Drives

Removing the Black Tires from the Large Driving Wheels and they become Pulley Wheels, and using some Rubber Bands to wrap over these Pulley Wheels, we can get a 4-Wheel Drive CarBot (see Fig. 5.4).

Fig. 5.4 Four-Wheel Drive CarBot using Pulley Wheels and Rubber Bands.

Adding “Spikes” to the Front Wheels would make this CarBot handle carpeted floors better or climb over small obstacles (see Fig. 5.5).

Fig. 5.5 Adding “Spikes” to Front Wheels.

Alternatively, adding “Spikes” to the “Spinning Top” robot would make it into a “Dancing Bear” or a “Waddling Penguin” (see Fig. 5.6).

Fig. 5.6 “Dancing Bear” / “Waddling Penguin”.

The PLAY700 kit does not actually have "gear" parts, but we can use the Large and Small Pulley Wheels and the Large Rubber Tires to build a Gear Train as shown in Fig. 5.7.

Fig. 5.7 Gear Train built from Pulley Wheels and Rubber Tires.

This Gear Train can be used to illustrate concepts such as Gear/Speed Ratio, Direction Reversal and Power Transmission. The Rubber Tire plays a very important role by its being "squeezed" at the contact points between adjacent Pulley Wheels, as this creates "friction" between the wheels which then allows rotational motions on one wheel to be transmitted to the other wheel (but in the reverse direction).

Ones can also use the Small Rubber Tire (http://www.robotis.us/small-tire-tr-37-1pcs/) to build an alternate Gear Train as shown in Fig. 5.8. The Small Rubber Tire is not included in the original PLAY700 kit (but it is included in the PLAY600 kit). In this configuration, ones can also use the Large Rubber Tire on the Large Pulley Wheel instead to achieve the same effect (but with less mechanical efficiency).

Fig. 5.8 Alternate Gear Train using Pulley Wheels and Small Rubber Tire.

Some readers probably already notice that if an identical gear train to the one shown in Fig. 5.7 is mounted on the other side of the CM-50, then we've got ourselves a 4-wheel drive as shown in Fig. 5.9. This version of 4-wheel drive would be more powerful and maneuverable than the one using rubber bands shown in Fig. 5.4, because the coupling via the pulley/tire interface is mechanically more efficient and the rubber bands can be too "flexible" to transmit high power needs. This "new" solution however requires the user to acquire two more Large Pulley Wheels and two more Large Black Tires beyond the parts that came with the original PLAY700 kit!

Fig. 5.9 4-WHEEL DRIVE CarBot.

Then, how about a Monster Truck version? (see Fig. 5.10), or a Rough Terrain version? (see Fig. 5.11). Please see Video 5.1 for a demonstration of these 3 ENHANCED CarBots.

Fig. 5.10 Monster Truck version.

Fig. 5.11 Rough Terrain version.

With the Monster Truck and Rough Terrain versions, we can now distinguish as separate the power transmission component (black pulley wheels and rubber tires) from the attached wheels themselves. Alas! These "Gear-based" 4-Wheel Drive, Monster Truck and Rough Terrain versions require more spare parts beyond a single PLAY700 kit!

5.2 Linkage Based Applications

5.2.1 Four-Legged DogBot

The original PLAY700 DogBot is essentially a CarBot (see Fig. 5.12), but it can be modified using a 4-bar Linkage system to become a 4-Legged DogBot as shown in Fig. 5.13. Each side of the DogBot's body uses an independent 4-bar Linkage system.

Fig. 5.12 Original Wheel-based PLAY700 DogBot.

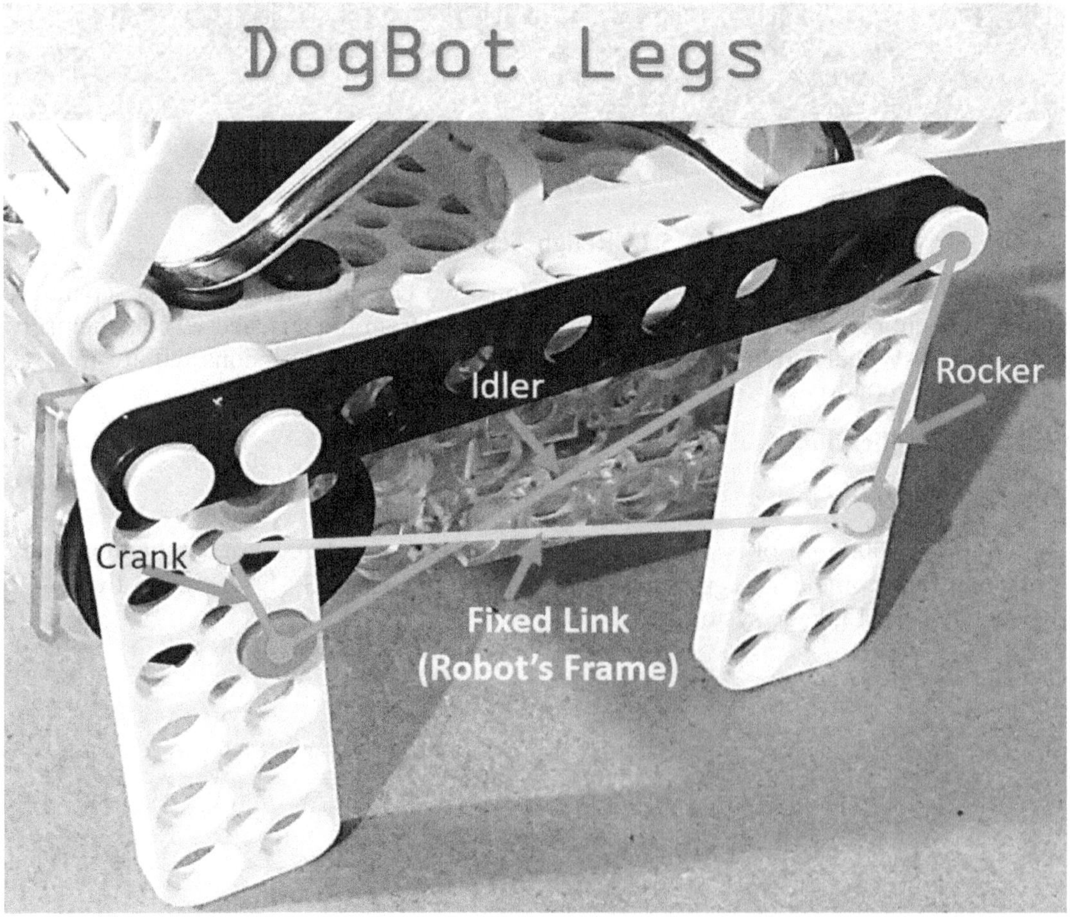

Fig. 5.13 4-Legged DogBot using 4-Bar Linkages.

Wikipedia has a compact write-up about 4-Bar linkages where the reader can learn more about this subject (https://en.wikipedia.org/wiki/Four-bar_linkage). The "Fixed Link" is usually the easiest one to identify as it is "fixed" on the robot frame and 1 of its 2 joint nodes is on the axle of the motor (see Fig. 5.13). This "motor" node is also part of the "Crank Link" which is usually identified with the motor's horn. The "Idler Link" could be the hardest to find as it does not always map to the obvious physical part that is used for it (see black linkage part in Fig. 5.13 and red linkage part in Fig. 5.14). The "Rocker Link" is usually the "leg" or "arm" for the robot, so it is relatively easy to identify.

5.2.2 Scorpion

The PLAY700 Scorpion robot has three 4-Bar linkages merged together to form 4-legs on each side of this robot (see Fig. 5.14):

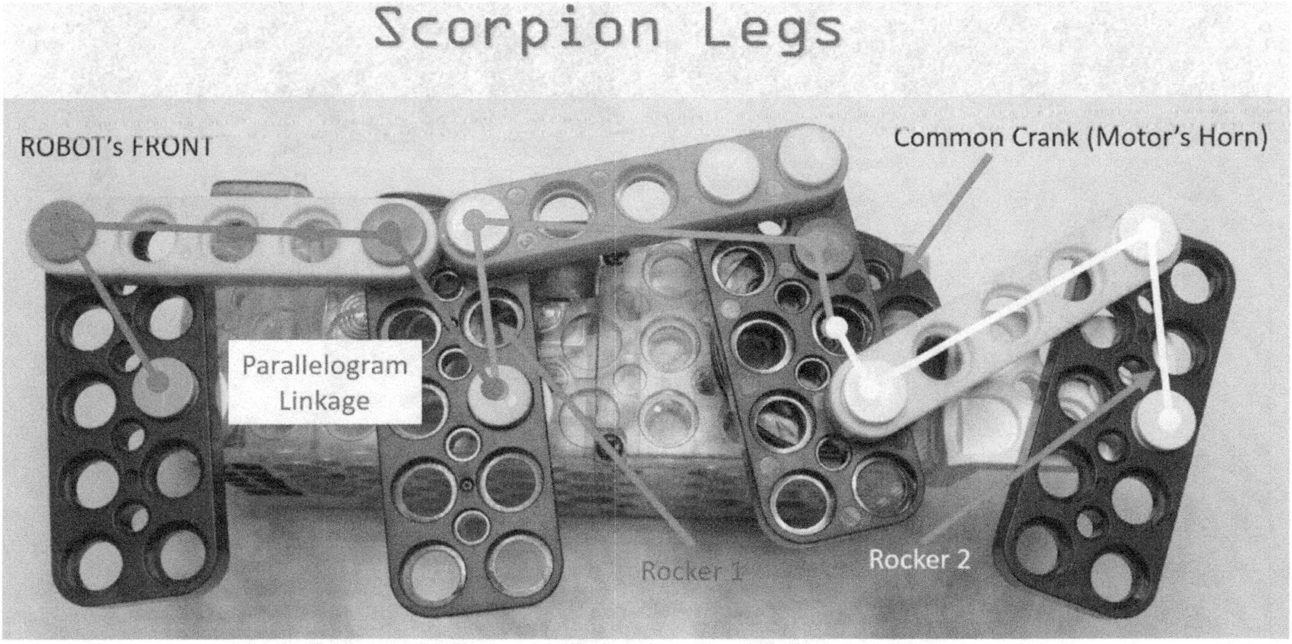

Fig. 5.14 Scorpion's Legs use three 4-Bar Linkages.

Going from the Robot's Front, the first 4-Bar linkage is a special one called "Parallelogram Linkage" (for obvious reasons). It is the Blue linkage.

The second 4-Bar linkage is colored Green. It is of the Crank-Rocker type and its Rocker is named Rocker 1 in Fig. 5.14.

The third 4-Bar linkage is also of the Crank-Rocker type, and it is colored Yellow. Its Rocker is labeled Rocker 2 in Fig. 5.14.

Although, in Kinematics sense, there are two separate Cranks - colored Green and Yellow - they are physically attached to the same physical part which is the Motor's Horn (the only "powered" link for both Crank-Rocker linkages).

5.2.3 Two-Arms Robot

A mixed "Gear/4-Bar Linkage" application is shown in Fig. 5.15 for a Two-Arms robot (nicknamed "Zombie").

Fig. 5.15 A Two-Arms robot - Zombie.

This design was chosen to get the 4-bar linkages higher and away from the ground surface and thus to clear them from potential interferences from low obstacles. See Video 5.2 for a demonstration of this robot.

5.2.4 Monkey's Crawling Motion

Figure 5.16 describes a Monkey robot that can crawl on all four, using its arms and legs. On each side of the Monkey's body, a typical Crank-Rocker mechanism is used. The Monkey's original design can be found in the R+Design Gallery for the PLAY503 kit (Monkey A).

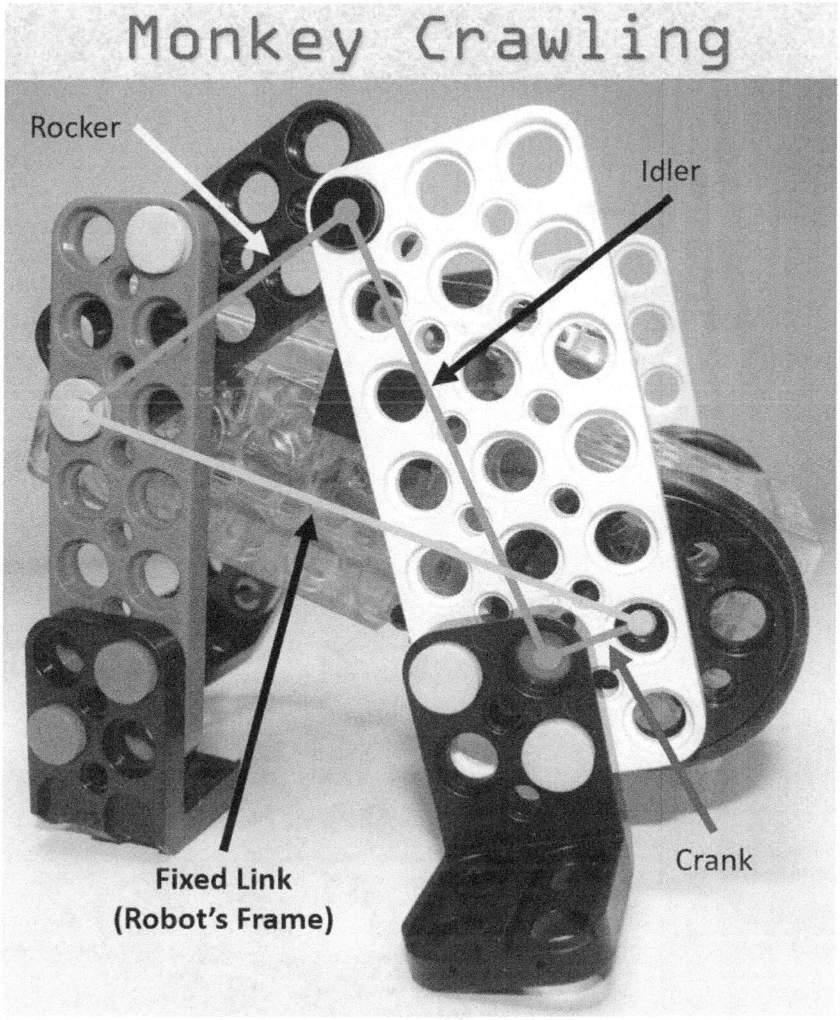

Fig. 5.16 Monkey Robot Crawling Mechanism.

See Video 5.3 for a demonstration of the motions that this Monkey robot can do.

5.2.5 Seal's Hitching Motion

Figure 5.17 describes a Seal robot that uses another "Crank-Rocker" mechanism for its flippers to reproduce the "hitching" maneuver used by real seals to move about on rocks and ice. Similar with the Monkey, each side of the Seal body has its own "Crank-Rocker" system, but this time, the Rocker is a 3-D structure that is "common" to both Left and Right 4-Bar Linkage systems. Because of this physical constraint, this Seal robot can only use its flippers to move forward or backward (i.e. synchronized turning out of the left and right motors), and it cannot turn left or right as these motions would require differential turning of the motors. The Seal's original design comes from the PLAY501 Kit (Seal B).

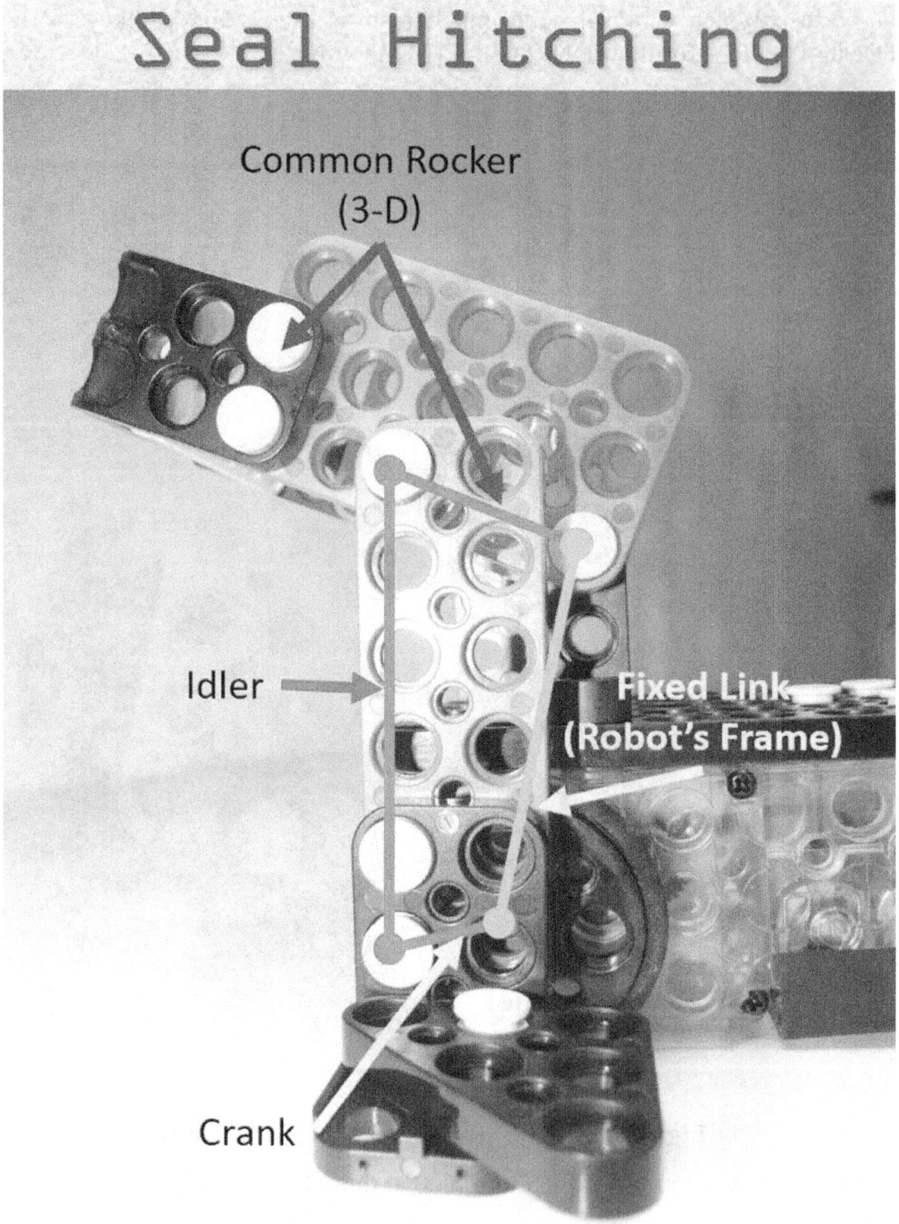

Fig. 5.17 Seal Robot Hitching Mechanism.

See Video 5.4 for a demonstration of the motions that this Seal robot can do. Although this video shows that the Seal lists a little bit to one side, this is because one motor can turn a little faster than the other one even though the TASK program commands them to turn at the same numerical speed setting, and friction effects may be different for each side's linkage system also.

5.2.8 Goldfish's Tail Flipping Motion

The Goldfish robot shares the same design approach as for the Seal because it also uses a common Rocker system that links the Left and Right sides of the robot (see Fig. 5.18). Interestingly, this common Rocker has several more components than the one for the Seal. The Goldfish's original design comes from the PLAY600 kit where two rivets are used to fasten the "black and white tail" part to the "yellow spine" component. The author prefers to use only one rivet for this function to provide more freedom (or slack if you will) to the tail's motions. See Video 5.5 for a demonstration of the motions that this Goldfish robot can do.

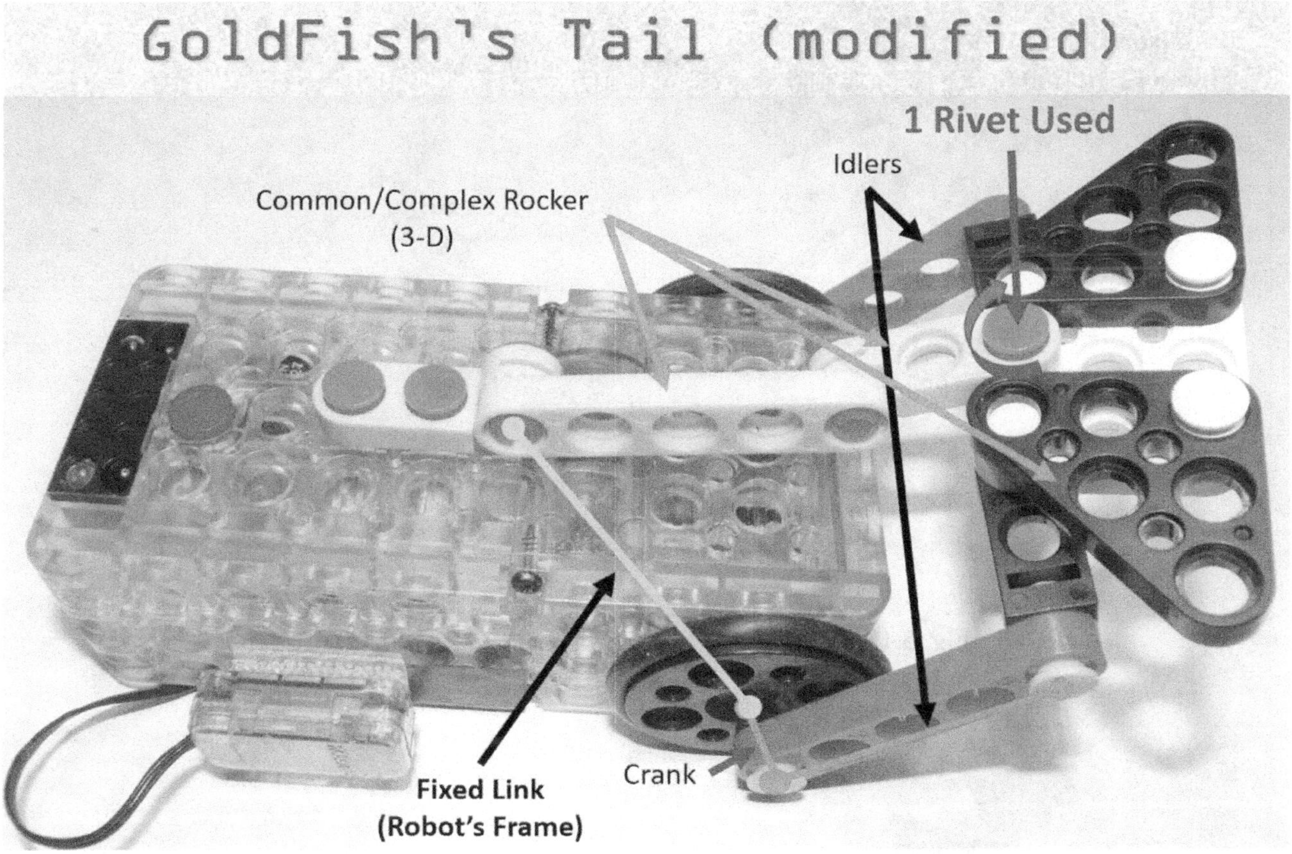

Fig. 5.18 Goldfish Robot Tail-Flipping Mechanism (modified).

5.2.9 Caterpillar's Crawling Motion

The Caterpillar robot is an even more interesting application of the 4-bar linkage concept (see Fig. 5.19). Normally, the Rocker component is connected to the robot's frame via a fixed joint/node as in previous robot designs. However, with the Caterpillar, the Rocker (i.e. yellow 5x9 plate) can "rock" and "slide" on the entire "top back edge" of the CM-50 controller as shown in Fig. 5.19. The original Caterpillar design comes from the OLLO K kit and uses two rivets for each left and right side of the "yellow-plate" rocker component (http://www.robotis.com/index/product.php?cate_code=10121111). This original design restricts the Caterpillar to forward and backward motions. The author prefers to use only one rivet for each side to provide the "yellow-plate" the ability to rotate left and right, thus allowing the whole Caterpillar to turn left and right quite easily (see Video 5.6 and Fig. 5.20 for a bottom view of the Caterpillar).

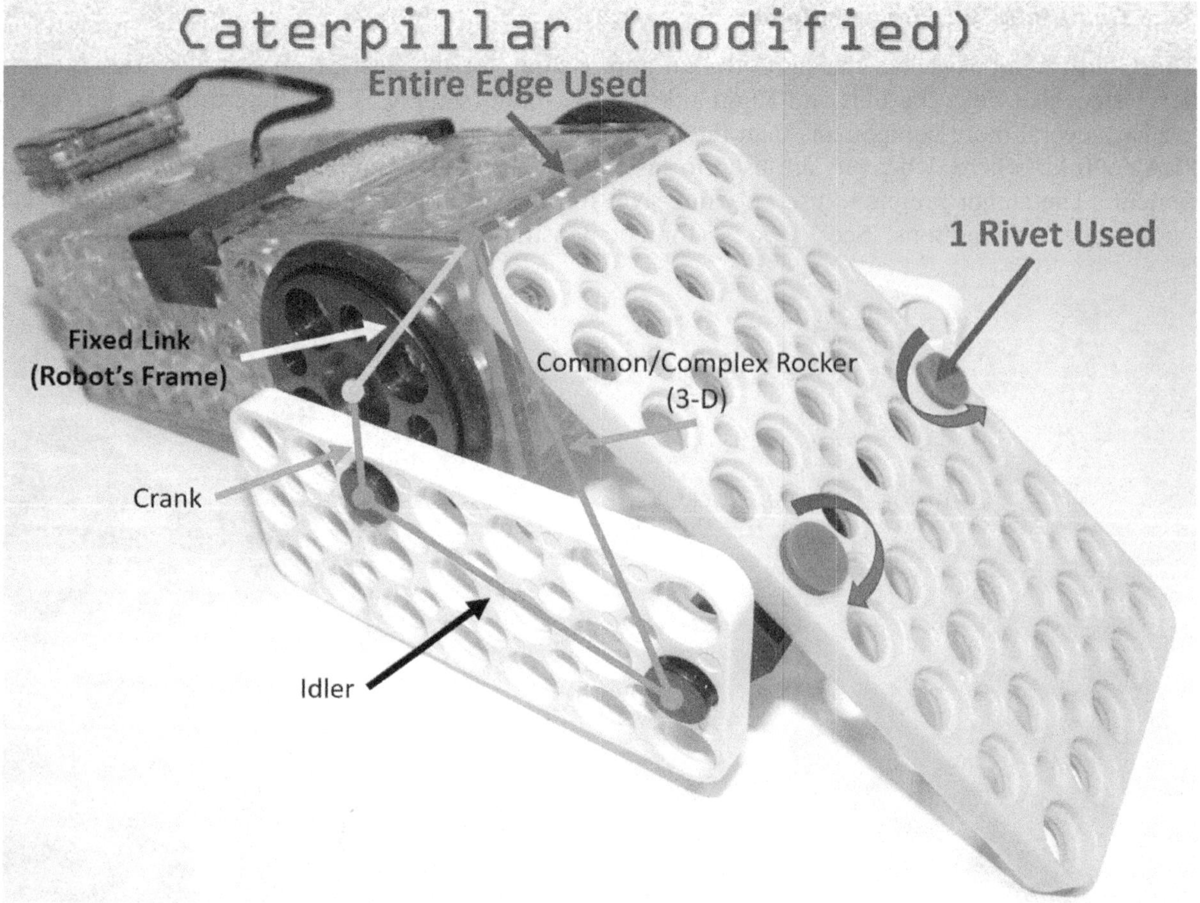

Fig. 5.19 Caterpillar Crawling Mechanism (modified).

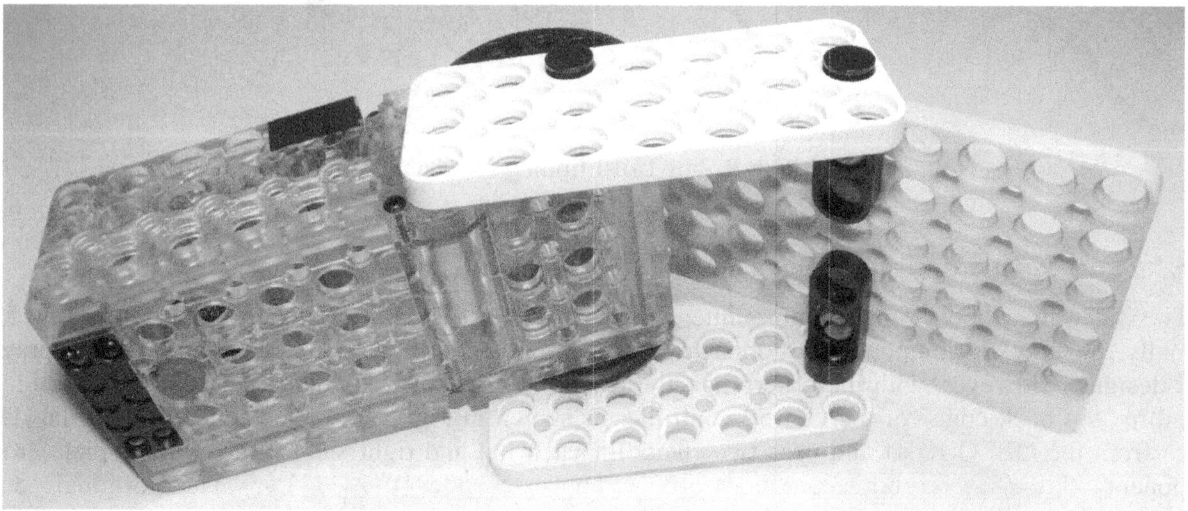

Fig. 5.20 Bottom View of Caterpillar Robot (modified).

Made in United States
North Haven, CT
05 November 2026

21765505R00089